at the
Clinical
Interface

Challenges at the Clinical Interface

CASE HISTORIES FOR CLINICAL BIOCHEMISTS

EDITED BY
Danielle B. Freedman, MBBS, FRCPath
James Hooper, MD, FRCPath
Philip J. Wood, BSc, MBA
David J. Worthington, PhD, FRCPath
Christopher P. Price, PhD, FRCPath

AACC Press

2101 L STREET, NW, SUITE 202
WASHINGTON, DC 20037-1558

©2001 American Association for Clinical Chemistry, Inc. All rights reserved. No part of this publication may be reproduced, stored in a retrieval systems, or transmitted in any form by electronic, mechanical, photocopying, or any other means without written permission of the publisher.

1 2 3 4 5 6 7 8 9 0   RRD   03 02 01

Printed in the United States of America

Library of Congress Cataloging-in-Publication Data

Challenges at the clinical interface: case histories for clinical biochemists / edited by Danielle Freedman...[et al.].
     p. ; cm.
  Includes index.
  ISBN 1-890883-52-2 (alk. paper)
  1. Clinical biochemistry—Case studies. I. Freedman, Danielle, 1953–
  [DNLM: 1. Chemistry, Clinical —Case Report. 2. Biochemistry—Case Report. QY 90 C437 2001]
  RB112.5.C465 2001
  616.07'56—dc21

2001037328

Contents

Preface _____ ix

Editors _____ xiii

Contributors _____ xiv

How to Use this Book _____ xx

1. Hypertension: the hidden cause _____ 1
2. A floppy infant _____ 7
3. Pregnancy, plus, plus _____ 11
4. A vomiting neonate _____ 15
5. A well-developed boy _____ 19
6. A gastrointestinal starting point _____ 23
7. Tired and weepy _____ 29
8. A surprising testosterone level _____ 35
9. Serum estradiol as a tumor marker: a cautionary tale _____ 41
10. A complicated pregnancy _____ 45
11. A complication of anticonvulsant therapy _____ 51
12. A complication of a road traffic accident _____ 57
13. Chronic hyponatremia _____ 61
14. A curious case of hyperkalemia _____ 65
15. Sickness in pregnancy _____ 69
16. Look before you leap _____ 73
17. A 52-y-old man with abdominal pain and vomiting _____ 77

18. An A to Z of intoxication — 83
19. Cushing's syndrome: one patient, two causes? — 89
20. An unusual CK-MB result — 93
21. Prolonged jaundice in a neonate — 97
22. Familial chronic fatigue — 103
23. A misleading case of abdominal pain — 109
24. Neonatal hypocalcemia — 113
25. Unusual biochemical changes after a flu-like illness — 119
26. A suspected overdose — 123
27. Prenatal screening and MoMs — 129
28. A difficult neurological case — 133
29. A case of abdominal pain and intermittent diarrhea — 137
30. Abdominal pain and vomiting in a 16-y-old girl — 143
31. Of course I am taking the tablets, doctor! — 147
32. A tale of two cations — 151
33. Hypocalcemia in the intensive-care unit — 155
34. Tale of the unexpected — 159
35. What a surprise! — 163
36. A child with constipation — 169
37. A woman eating tuna — 175
38. There's a gap: anion, osmolar, or what? — 179
39. Long walks and dark urine — 185
40. Drowsiness and confusion in an 83-y-old woman — 191
41. Sugar, water—and more — 195
42. Muscle pain and hypolipidemic therapy — 201
43. A young psychotic long-distance traveler — 207

44. Convulsions in a neonate	213
45. Found in a coma	217
46. Metabolic encephalopathy	221
47. A diabetic emergency	225
48. An uncommon urine dipstick reaction	231
Reference ranges for analytes in blood (adults)	235
Reference ranges for analytes in blood (infants)	239
Reference ranges for analytes in urine (adults)	240
Reference ranges for miscellaneous biochemical parameters	241
Reference ranges for therapeutic drug concentrations (serum concentrations)	241
Reference ranges for hematological parameters (adults)	242
Cases by broad subject area	243
Index	245

Preface

Laboratory medicine plays a vital role in the detection, diagnosis, and management of disease. There are many facets to this role, of which the provision of analytical results and interpretation of the data are probably the most public face of the service. There are many other activities that provide the foundations of laboratory medicine, including the basic research generating the knowledge base; design of reagents, devices, and equipment; education and training of both users and laboratory professionals; quality assurance; audit; and continuing education. One of the most challenging aspects in the training of laboratory medicine professionals is in communication, attaining the knowledge and skill to ensure that the best use is made of the service by clinicians.

Laboratory medicine professionals who are trained to provide the interpretative aspects of the service operate at an interface between the ward, clinic, or primary care office and the laboratory. This pivotal position has not always been the case; many of the early diagnostic investigations were undertaken at the bedside or in a ward side room. The separation that has arisen between the clinical area and the laboratory is a consequence of several factors, including the repertoire and workload of investigations requested, the evolution of sophisticated analytical technology to meet these needs, health and safety considerations, and economic pressures to achieve economies of scale. The latter considerations have taken matters further in some places, with a significant part of the laboratory activity becoming physically divorced from the clinical environment. This demarcation or compartmentalization has extended into other aspects of the provision of health care, including subspecialization in professional practice, resource allocation, and ultimately experience.

As a consequence, new challenges arise in that individuals practicing in specialized areas of health care have to interact closely to provide the required integration of care. This occurs extremely successfully in the multidisciplinary teams working on the wards, in the clinics, and at the health centers. However, for some of the reasons identified earlier, the laboratory specialist often remains physically isolated—with the risk of being professionally and intellectually isolated as well.

What is the effect of specialization on education, training, and professional practice? In many disciplines of laboratory medicine the interface between the clinical environment and the laboratory is provided by either medical or science graduates (pathologists, clinical chemists, etc.); the proportion of science graduates varies according to specialty and by country. However, the fundamental requirements are the same—put very simply, a combination of "clinical knowledge and experience" and "scientific knowledge and experience." It is fairly easy to describe scientific knowledge and experience in relation to the practice of laboratory medicine—the pathophysiology of the disease, origin of the marker (test), analytical technology, biological variation and statistics, quality assurance, etc. It is a little more difficult to describe clinical knowledge and experience in relation to the practice of laboratory medicine. In the case of the qualified doctor, it is embodied in his or her training and registration to practice. In the case of the clinical scientist (the individual whose primary degree is in science rather than medicine), the situation is more complex. It is embodied in many successful training programs—often undertaken in parallel with medical graduates, and in the practicing "products" of these programs. Inherent in all of these training schemes is in-service training or on-the-job training and attendance at ward rounds and other clinical meetings, together with the examinations associated with higher specialist training and registration to practice. An implicit part of all of these training activities is the study (and sometimes presentation) of clinical cases. The point was made a few years ago in an introduction to a set of clinical cases (Hooper J. Cases in clinical biochemistry. J R Soc Med 1996;89:213P) that said "Clinical cases remind us of the relevance of laboratory investigations in diagnosis, treatment and monitoring of patients, but more importantly are memorable and enable us to build up our own mental picture gallery of our special subject for easy reference."

Individual case reports have always been crucial to the understanding of the pathophysiology of many diseases, especially at a fundamental level, but these are rare and exceptional. Clinical cases also have an important educational role and one that is increasingly seen as a means of imparting the relevance of a clinical subject to students. Indeed, with the emphasis on problem-based learning by many educational authorities, clinical cases provide good examples of how to solve real clinical problems. They also form part of the approach to integrated education with less reliance on the traditional divide between preclinical and clinical studies.

Clinical cases provide insight into the diagnostic process and allow students to develop their interpretative skills in a real-life situation, but in a protected, educational setting. The cases call upon students to use their imagination to synthesize aspects of the clinical presentation and investigation into a plausible set of possible diagnoses. Whether aware or not, the students use the principle of Occam's razor to provide the most simple (and therefore the most likely) explanation for the presenting symptoms, signs and investigations. [Occam's razor is a rule of thumb named for its formulator, William of Occam (died 1349): When explaining something, assumptions must not be needlessly multiplied (the principle of economy of explanation)].

This book sets out to meet this educational need by presenting several cases, all from the time of presentation of the patient to the achievement of a diagnosis and establishment of an appropriate intervention strategy. These are real cases, reported as far as possible in the way that they unfolded. This faithfulness to what really happened is an important feature of these case histories because it helps the reader to understand and appreciate the process of clinical practice.

Twelve years ago, the editors of this book initiated a series of continuing education meetings in clinical biochemistry. The core of the meetings was a small number of invited speakers who discussed topics of current interest, with the rule being that more time was given over to debate than didactic presentation. One of the sessions was dedicated to the presentation of clinical cases by members of the audience—again, more discussion than presentation. These cases proved to be one of the most popular features of the meeting, probably because they:

- Provided a holistic view of the patients' problems that most laboratory professionals did not always have the opportunity to see on a day-to-day basis

- Provided insight into some unusual cases

- Enhanced practitioners' clinical skills

- Enhanced practitioners' experience

A selection of these cases is presented in this book; in the main, they are a collection of more unusual cases, although some basic clinical problems are also included.

Acknowledgements

The participants at the continuing education meetings over the years—including the editors—have always spoken extremely highly of the clinical case sessions. We felt that this benefit could be shared with a much wider community of laboratory and clinical colleagues.

This book of cases would not have been possible, however, without the contribution of the participants at these meetings. We would like to express our heartfelt gratitude to colleagues and friends who attended, and in particular to Instrumentation Laboratory (UK) Ltd, which sponsored the meetings and undertook most of the logistics work for them.

We would also like to express our particular gratitude to all of the contributors to the case presentations who are recognized elsewhere in this book.

Finally, we are indebted to Laura Roberts, who provided invaluable support formatting all of the cases and ensuring that both authors and editors met the required deadlines. We are also extremely grateful for the support we have received from Joanna Grimes and colleagues at AACC.

Danielle B. Freedman
James Hooper
Philip J. Wood
David J. Worthington
Christopher P. Price

June 2001

Editors

Danielle B. Freedman, MBBS, FRCPath
Department of Chemical Pathology
Luton & Dunstable Hospital
Luton LU4 0DZ
United Kingdom

James Hooper, MD, FRCPath
Department of Clinical Biochemistry
Royal Brompton Hospital
Sydney Street
London SW3 6NP
United Kingdom

Philip J. Wood, BSc, MBA
Instrumentation Laboratory UK Ltd
Kelvin Close
Birchwood Science Park
Warrington
Cheshire WA3 7PB
United Kingdom

David J. Worthington, PhD, FRCPath
Clinical Chemistry Department
Birmingham Children's Hospital
Steelhouse Lane
Birmingham B4 6NH
United Kingdom

Christopher P. Price, PhD, FRCPath
Department of Clinical Biochemistry
Barts and the London NHS Trust
The Royal London Hospital
London E1 1BB
United Kingdom

Contributors

Julian Barth, MD, MRCPath
Department of Clinical Biochemistry & Immunology
Leeds General Infirmary
Leeds LS1 3EX
United Kingdom

Graham Bayly, MRCP, MRCPath
Clinical Biochemistry Department
Bristol Royal Infirmary
Bristol BS2 8HW
United Kingdom

Geoffrey Beckett, PhD, FRCPath
Department of Clinical Biochemistry
The Royal Infirmary
Lauriston Place
Edinburgh EH3 9YW
United Kingdom

Michelle Bignell, PhD, DipRCPath
Birmingham Children's Hospital
Steelhouse Lane
Birmingham B4 6NH
United Kingdom

Jacky Burrin, PhD, FRCPath
Molecular Endocrinology Laboratory
St Bartholomew's Hospital
West Smithfield
London EC1A 7BE
United Kingdom

Jose Cabrera-Abreu, LMS, MRCPath
Department of Clinical Chemistry
Birmingham Children's Hospital
Steelhouse Lane
Birmingham B4 6NH
United Kingdom

Ceridwen Dawkins, MB ChB, FRCPath
Department of Chemical Pathology
Frenchay Hospital
Bristol BS16 1LE
United Kingdom

Michael Diver, PhD, MRSC
Department of Clinical Chemistry
Royal Liverpool University Hospital
Prescot Street
Liverpool L7 8XP
United Kingdom

George Elder, MD, FRCPath
Medical Biochemistry Department
University of Wales College of Medicine
Heath Park
Cardiff CF14 4XN
United Kingdom

Danielle B. Freedman, MBBS, FRCPath
Department of Chemical Pathology
Luton & Dunstable Hospital
Luton LU4 0DZ
United Kingdom

Peter Gosling, PhD, FRCPath
Department of Clinical Biochemistry
Selly Oak Hospital
Raddlebarn Lane
Birmingham B29 6JD
United Kingdom

Trevor Gray, FRCP, FRCPath
Department of Clinical Chemistry
Northern General Hospital NHS Trust
Herries Road
Sheffield S5 7AU
United Kingdom

Ian Gunn, MB ChB, FRCPath
Biochemistry Department
Law Hospital
Carluke
Lanarkshire ML8 5ER
United Kingdom

Michael Hallworth, MA, FRCPath
Department of Clinical Biochemistry
Royal Shrewsbury Hospital
Mytton Oak Road
Shrewsbury SY3 8XQ
United Kingdom

James Hooper, MD, FRCPath
Department of Clinical Biochemistry
Royal Brompton Hospital
Sydney Street
London SW3 6NP
United Kingdom

Garry John, PhD, FRCPath
Department of Clinical Biochemistry
Barts and the London NHS Trust
St Bartholomew's Hospital
London EC1A 7BE
United Kingdom

Jonathan Kay, MBBS, FRCPath
Department of Clinical Biochemistry
John Radcliffe Hospital
Oxford OX3 9DU
United Kingdom

Edmund Lamb, PhD, MRCPath
Department of Clinical Biochemistry
Kent & Canterbury Hospital NHS Trust
Ethelbert Road
Canterbury CT1 3NG
United Kingdom

John Land, BM BCh, MRCPath
Department of Clinical Biochemistry
National Hospital for Neurology
Queen Square
London WC1N 3BG
United Kingdom

Ruth Lapworth, BSc, FRCPath
Department of Clinical Biochemistry
William Harvey Hospital
Ashford
Kent TN24 0LZ
United Kingdom

Steven C. Martin, MB ChB, MRCPath
Department of Clinical Biochemistry
West Suffolk Hospital
Bury St Edmunds IP33 2QZ
United Kingdom

Paul Masters, MB ChB, MRCPath
Department of Chemical Pathology
Chesterfield & N Derbyshire Royal Hospital
Calow
Chesterfield S44 5BL
United Kingdom

Gwyn McCreanor, PhD, MRCPath
Biochemistry Department
Kettering General Hospital
Rothwell Road
Kettering
Northants NN16 8UZ
United Kingdom

Martin Myers, PhD, MRCPath
Department of Clinical Chemistry
Royal Preston Hospital
Sharoe Green Lane
Preston PR2 4HG
United Kingdom

Kate Noonan, BSc, MSc
Department of Clinical Biochemistry
Barts and the London NHS Trust
The Royal London Hospital
London E1 1BB
United Kingdom

John O'Connor, PhD, MRCPath
Department of Biochemistry
Eastbourne District General Hospital
Kings Drive
Eastbourne BN21 2UD
United Kingdom

Michael Penney, MD, FRCPath
Department of Chemical Pathology
Royal Gwent Hospital
Newport
Gwent NP20 2UB
United Kingdom

Christopher P. Price, PhD, FRCPath
Department of Clinical Biochemistry
Barts and the London NHS Trust
The Royal London Hospital
London E1 1BB
United Kingdom

Laurence Robinson, MSc, FRCPath
Department of Chemical Pathology
Wythenshawe Hospital
South Manchester University Hospitals
Manchester M23 9LT
United Kingdom

Colin Samuell, MIBiol, FRCPath
Chemical Pathology
UCL Hospitals
Windeyer Building
Cleveland Street
London W1P 6DB
United Kingdom

Brian Senior, PhD, CMS
Department of Clinical Chemistry
Royal Bolton Hospital
Minerva Road
Bolton BL4 0JR
United Kingdom

Janet Smith, MSc, FRCPath
Department of Clinical Biochemistry
University Hospital Birmingham NHS Trust
Selly Oak Hospital
Raddlebarn Lane
Birmingham B29 6JD
United Kingdom

Geoffrey Smith, MB ChB (Edin), MRCPath
Southern Community Laboratories Ltd
444 Durham Street North
Christchurch
PO Box 21 049
New Zealand

Catherine Street, BSc, PhD
Department of Clinical Biochemistry
Barts and the London NHS Trust
The Royal London Hospital
London E1 1BB
United Kingdom

Peter M. Timms, BSc, MRCPath
Department of Clinical Biochemistry
Barts and the London NHS Trust
St Bartholomew's Hospital
West Smithfield
London EC1A 7BE
United Kingdom

Ian Watson, PhD, MRCPath
Department of Clinical Biochemistry
University Hospital Aintree
Longmoor Lane
Liverpool L9 7AL
United Kingdom

Hazel Wilkinson, MB BS, FRCPath
Biochemistry Department
York District Hospital
Wigginton Road
York YO3 7HE
United Kingdom

David J. Worthington, PhD, FRCPath
Clinical Chemistry Department
Birmingham Children's Hospital
Steelhouse Lane
Birmingham B4 6NH
United Kingdom

How to Use this Book

All of the cases are presented in a common format, and we have attempted to use an interactive style that challenges the reader's knowledge in a way that might pertain in routine practice. Thus, the first page of each case sets out the presenting symptoms and the information gleaned from the initial consultation. The reader is then asked to consider provisional diagnoses and the investigations that are likely to be requested.

On the second page the answers to the questions are provided, including the results of investigations actually performed. In addition, there may be additional information given, such as the course of the patient's stay or the patient's symptoms. The reader is then asked whether he or she wishes to modify the provisional diagnoses and whether further investigations are indicated. The reader is encouraged to mask the third page (on the right hand side) with a bookmark or piece of paper to hide the results while attempting to answer the questions.

The remaining sections of each case then deal with the course of the patient's management and condition and the final diagnosis. There then follows a section that sets out the key features of the disease and its diagnosis and treatment. Finally, some important learning points are identified, together with some references for further reading.

All of the cases are supplied with results in both conventional and Système International (SI) units. The conversions may not always appear to be exact; this inexactitude reflects the fact that all of the cases derived from the United Kingdom where SI units are used. In undertaking the conversion to conventional units, the numbers have been rounded up or down to the appropriate number of significant figures. It is perhaps also worth mentioning that urea is reported in numerical terms as blood urea nitrogen for conventional units and in "whole molecule" molar terms for SI units in order to ensure familiarity with the numerical results.

A table of all of the relevant reference ranges in adults is supplied—again, in conventional and SI units. In some cases the results have been 'normalized' to a given reference range from the original patient records to avoid the need to provide laboratory-specific reference ranges (as, for example, in the case of amylase and alkaline phosphatase). Pediatric reference ranges are supplied in the text where appropriate.

1. Hypertension: the hidden cause

PRESENTATION

HISTORY OF PRESENTING COMPLAINT
- A 75-y-old man was referred to the outpatient clinic. He had had hypertension for 20 y, initially controlled on Slow Trasicor, co-amilozide, and nifedipine
- His hypertension was now difficult to control on nifedipine and enalapril
- His blood pressure was 180/100 mmHg
- For the past 4 mo his serum potassium had been between 3.2 and 3.6 mEq/L (3.2 and 3.6 mmol/L)

PAST MEDICAL HISTORY
- He had had a mild cerebrovascular accident and a cataract operation

SOCIAL HISTORY
- He had lived in the United Kingdon since 1947, but was originally from Italy
- He stopped smoking in 1955 and only drank alcohol occasionally
- He was a retired engineer

FAMILY HISTORY
- His mother died in her 40s, but the cause of her death was unknown

SUBSEQUENT QUESTIONING
- Unremarkable

ON EXAMINATION
- He was overweight with a body mass index of 28.5. He was clinically euthyroid
- Blood pressure was 150/100 mmHg, heart sounds were normal, and there were no other abnormalities found

---

WHAT IS YOUR PROVISIONAL DIAGNOSIS?

WHAT INVESTIGATIONS WOULD YOU REQUEST?

PROVISIONAL DIAGNOSIS
- Primary hypertension, idiopathic or essential hypertension (95% of cases)
- Secondary hypertension (5% of cases)
  - Renal or renovascular
  - Endocrine (pheochromocytoma, Conn's syndrome, acromegaly, Cushing's syndrome, primary hyperparathyroidism, thyrotoxicosis)
  - Coarctation of the aorta
  - Iatrogenic—for example, caused by nonsteroidal anti-inflammatory drugs or oral contraceptives

INITIAL INVESTIGATIONS

Serum
Sodium	143 mEq/L	(143 mmol/L)
Potassium	3.5 mEq/L	(3.5 mmol/L)
Urea	14 mg/dL	(5.0 mmol/L)
Cholesterol	158 mg/dL	(4.1 mmol/L)
Glucose	86 mg/dL	(4.8 mmol/L)

24-h urine
Sodium	120 mEq	(120 mmol)
Potassium	53 mEq	(53 mmol)

Note: The patient was on enalapril, an angiotensin-converting enzyme (ACE) inhibitor, and a repeat plasma potassium was 3.0 mEq/L (3.0 mmol/L)

WHAT FURTHER INVESTIGATIONS, IF ANY, WOULD YOU REQUEST?

HAVE YOU MADE ANY CHANGES TO YOUR PROVISIONAL DIAGNOSIS?

WHAT IS YOUR DIAGNOSIS?

WORKING DIAGNOSIS
- Hypertension with persistent hypokalemia

FURTHER INVESTIGATIONS
Patient admitted for renin and aldosterone measurements

Patient preparation:
- Investigations should only be undertaken in hypertensive patients with persistent hypokalemia [potassium <3.7 mEq/L (<3.7 mmol/L) on at least three occasions]
- There should be an inappropriate loss of potassium in urine [potassium >35 mEq/24 h (>35 mmol/24 h)]
- If patient on:
    Diuretics
    Nifedipine
    ACE inhibitors
    Spironolactone
    β-Blockers
  These must be discontinued for 1–2 wk before test. Treat hypertension with α-blocker, such as doxazosin
- Give potassium supplements to render patient normokalemic

Remember, gross potassium depletion inhibits aldosterone production, which may lead to a normal value in patients with Conn's syndrome (primary hyperaldosteronism)

RESULTS
24-h urine
  Sodium   206 mEq   (206 mmol)
  Potassium   105 mEq   (105 mmol)
Serum
  Sodium   141 mEq/L   (141 mmol/L)
  Potassium   3.2 mEq/L   (3.2 mmol/L)
  Creatinine   1.1 mg/dL   (96 μmol/L)
  Urea   9 mg/dL   (3.1 mmol/L)
Plasma renin activity
  0800 h   <0.24 ng/h/mL   (<0.2 pmol/h/mL)
  0830 h   <0.24 ng/h/mL   (<0.2 pmol/h/mL)
Aldosterone
  0800 h   227 pg/mL   (630 pmol/L)
  1200 h   140 pg/mL   (390 pmol/L)

PROGRESS
Patient underwent a magnetic resonance imaging scan, which showed a small adenoma in the right adrenal gland

FINAL DIAGNOSIS: Primary hyperaldosteronism (Conn's syndrome)

MANAGEMENT

Patient was treated with spironolactone, nifedipine, and aspirin, and his blood pressure came back to normal. He was maintained on these, and his potassium was consistently between 4.0 and 4.8 mEq/L (4.0 and 4.8 mmol/L)

KEY FEATURES

PRESENTATION
- Uncontrolled hypertension with persistent hypokalemia
- Hypokalemia persists despite being on an ACE inhibitor

BIOCHEMISTRY

FIGURE THE RENIN-ANGIOTENSIN SYSTEM

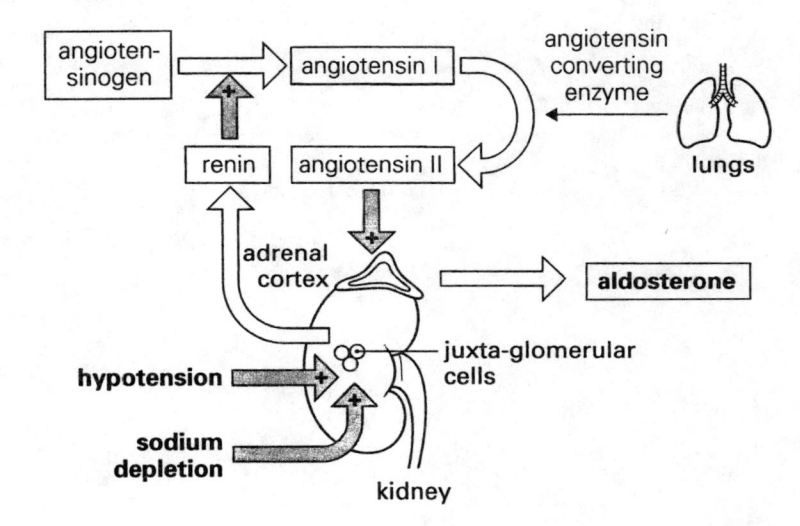

- Diagnosis of primary hyperaldosteronism in a hypertensive patient depends on plasma potassium, renin, and aldosterone measurements. A suppressed, nonstimulatable plasma renin activity has to be demonstrated
- Other causes of hypertension must be considered, including primary (90–95% of cases) or secondary (5% of cases) including renal or renovascular disease
    - Elementary biochemical tests are usually sufficient to exclude most renal and endocrine causes of hypertension
    - Young patients with high blood pressure merit more detailed investigation
    - The aim of the laboratory investigation of hypertension is:
        - To exclude rare identifiable renal or adrenal causes
        - To help detect the evidence of target organ damage, such as renal impairment
        - To detect elevation of serum lipids or glucose intolerance because each increases cardiovascular risk and influences choice of antihypertensive agent
        - To detect renal disease or suggest diabetes mellitus with the aid of proteinuria, hematuria, and glycosuria

- First-line biochemical tests
  - Sodium and potassium: In primary hyperaldosteronism, the sodium tends to be toward the upper limit of the normal range. In secondary hyperaldosteronism due to malignant hypertension or renal or renovascular disease, the sodium will be low
  - Marked hypokalemia occurs in patients with Conn's syndrome, but note that diuretics are the commonest cause of hypokalemia, and these drugs must be stopped for at least 2 wk before serum potassium measurement is undertaken
  - Hypokalemia is usually associated with metabolic alkalosis and a high serum bicarbonate concentration
- Serum potassium concentrations raised in acute renal failure and also possibly in patients who are on potassium supplements or potassium-sparing diuretics and can be found in patients on ACE inhibitors, such as captopril and enalapril
- Serum creatinine will assess renal function
- Serum calcium can be useful. Hypertension is found in 50% of patients with primary hyperparathyroidism. The mechanism is uncertain, but note that thiazide diuretics may rarely cause hypercalcemia and should be stopped for a few weeks before rechecking the calcium
- Serum uric acid is raised in ~40% of patients with essential (idiopathic) hypertension and more common in those with renal failure. In a patient with unexplained hyperuricemia, history may indicate excessive intake of alcohol
- Liver function tests (γ-glutamyltransferase) may indicate excess alcohol as a cause of hypertension
- Glucose (random or postprandial) can be useful. It may be necessary to use a fasting glucose estimation. About 50% of people with type 1 or type 2 diabetes have hypertension, and as many as 10% of people with hypertension also have diabetes
- Fasting lipids should be measured as part of the cardiovascular risk profile and also to influence choice of antihypertensive agents. Note that β-blockers and thiazide diuretics adversely affect plasma lipid profile

MANAGEMENT
- Depends on the cause, which may be adenoma, hyperplasia, or carcinoma (rare). In the case of adenoma and carcinoma, removal of the adrenal gland if the patient is fit for surgery. For hyperplasia or when the patient is unfit for surgery, use of aldosterone antagonists—namely, spironolactone—possibly together with other antihypertensives

POINTS TO REMEMBER
- The British Hypertension Society Guidelines recommend that physicians aim for a diastolic blood pressure of <85 mmHg and systolic blood pressure of <140 mmHg (British Hypertension Society Guidelines for Management of Hypertension. Report of the Third Working Party of the British Hypertension Society. J Hum Hypertens 1999;13:569–92.)
- Underlying causes of hypertension are found in <5% of the population, but they may be surgically remediable
- Think about primary hyperaldosteronism as one of the few curative causes of hypertension
- Suspect primary hyperaldosteronism (Conn's syndrome) in patients who have persistent hypertension that is difficult to control and persistent hypokalemia who are not on diuretics; patients may have associated metabolic alkalosis. Sodium may be normal or toward the upper end of the reference range
- Ensure the full protocol is followed for renin and aldosterone, including that the patient must be normokalemic at the time of investigation

- Plasma renin activity will be suppressed even on ambulation with an elevated aldosterone, both recumbent and ambulant

REFERENCES
1. Elliott H. Hypertension: investigation and management. Medicine 1998;26:141–5.
2. Gomez-Sanchez CE. Primary aldosteronism and its variants. Cardiovasc Res 1998;37:8–13.
3. Harvey JM, Vivas DG. Biochemical investigation of hypertension. Ann Clin Biochem 1990;27: 287–96.

THIS CASE WAS PRESENTED BY DR. DANIELLE FREEDMAN, LUTON AND DUNSTABLE HOSPITAL, LUTON, UK

2. A floppy infant

PRESENTATION

HISTORY OF PRESENTING COMPLAINT
- A 23-mo-old Asian girl presented with hypotonia, deafness, developmental delay, and a cardiac conduction defect

PAST MEDICAL HISTORY
- Normal delivery at term
- Poor feeder
- First presented at 4 mo with hypotonia and plasma lactate concentrations of 32 mg/dL (3.6 mmol/L)
- Despite extensive investigations, no inherited metabolic defects had been diagnosed

SOCIAL HISTORY
- First child of consanguineous parents (first cousins)

FAMILY HISTORY
- No family history of note

MEDICATION
- None

ON EXAMINATION
- Floppy
- Edematous

WHAT IS YOUR PROVISIONAL DIAGNOSIS?

WHAT INVESTIGATIONS WOULD YOU REQUEST?

PROVISIONAL DIAGNOSES
- Congestive cardiac failure
- Renal failure

INITIAL INVESTIGATIONS
Serum

Sodium	125 mEq/L	(125 mmol/L)
Potassium	3.6 mEq/L	(3.6 mmol/L)
Urea	35 mg/dL	(12.4 mmol/L)
Creatinine	1.1 mg/dL	(96 µmol/L)
Chloride	80 mEq/L	(80 mmol/L)
Bicarbonate	27 mEq/L	(27 mmol/L)
Osmolality	276 mOsm/kg	(276 mOsm/kg)
Albumin	3.7 g/dL	(37 g/L)
Glucose	92 mg/dL	(5.1 mmol/L)
Lactate	64 mg/dL	(7.1 mmol/L)
Aldosterone	>120 pg/dL	(>3300 pmol/L)

Plasma

Renin activity	23.6 ng/h/mL	(32.2 pmol/h/mL)

WHAT FURTHER INVESTIGATIONS, IF ANY, WOULD YOU REQUEST?

HAVE YOU MADE ANY CHANGES TO YOUR PROVISIONAL DIAGNOSIS?

WHAT IS YOUR DIAGNOSIS?

WORKING DIAGNOSIS
- Secondary hyperaldosteronism due to congestive cardiac failure

MANAGEMENT
- She was initially treated with diuretics to reduce the edema, but died shortly afterwards
- Cardiac ultrasound showed an enlarged heart

KEY FEATURES

PRESENTATION
- The cardiac conduction defect prevented the heart from contracting in a properly synchronized fashion (DiGeorge syndrome); hence, there was a reduction in the amount of blood effectively circulating. This resulted in reduced clearance of interstitial fluid into the veins and back to the heart and pooling of fluid in the periphery (edema)

BIOCHEMISTRY
- The elevated renin and aldosterone concentrations are due to the reduced circulating blood volume, which stimulates the juxtaglomerular apparatus of the kidney. This activation of the renin-angiotensin system results in aldosterone secretion and causes increased sodium reabsorption in the kidney. The final result is an increase in the circulating plasma volume

DIAGNOSIS
- Secondary hyperaldosteronism in response to congestive cardiac failure and a reduced circulating blood volume

POINTS TO REMEMBER
- Be aware of the causes of biochemical abnormalities
- Understand how to classify the causes of hyponatremia
- Know how to use laboratory investigations to diagnose the cause of hyponatremia
- Understand the physiology of the renin-aldosterone system
- Understand the pathophysiology of hyperaldosteronism

REFERENCES
1. Levine LS, DiGeorge AM. Excess mineralocorticoid secretion. In: Behrman RE, Kliegman RM, Jenson HB, eds. Nelson textbook of pediatrics, 16th ed. Philadelphia: WB Saunders, 2000:1739–41.

THIS CASE WAS PRESENTED BY DR. STEVEN C. MARTIN, WEST SUFFOLK HOSPITAL, BURY ST EDMUNDS, UK

3. Pregnancy, plus, plus

PRESENTATION

HISTORY OF PRESENTING COMPLAINT
- A 25-y-old primiparous woman presented at ~24 wk gestation to her primary-care physician with excessive weight gain and acne

PAST MEDICAL HISTORY
- The patient had been well before and during her pregnancy

DRUG HISTORY
- She had stopped taking the oral contraceptive pill shortly before conceiving and had not been receiving any other medication

ON EXAMINATION
- She had severe facial and truncal acne
- There were marked purple striae over her breasts, abdomen, buttocks, and thighs
- Her blood pressure was elevated at 145/85 mmHg, having been normal (130/70 mmHg) at her antenatal booking appointment

INITIAL INVESTIGATIONS
- Serum electrolytes, glucose, liver function tests, urate, full blood count, platelets, and urinary 4-hydroxy-3-methoxymandelic acid (HMMA) were all normal
- Urine total protein was increased (0.22 g/24 h) and urinary free cortisol was elevated [258 µg/24 h (712 nmol/24 h)] compared with the nonpregnant reference range

---

WHAT IS YOUR PROVISIONAL DIAGNOSIS ?

WHAT INVESTIGATIONS WOULD YOU REQUEST?

PROVISIONAL DIAGNOSIS
- Pre-eclampsia and Cushing's syndrome

FURTHER INVESTIGATIONS
- The patient was admitted for bed rest and further investigations. An ultrasound examination revealed normal fetal growth and a 5-cm mass in the patient's right adrenal gland
- Blood samples were collected to assess her diurnal variation in cortisol concentration, for basal adrenocorticotrophic hormone (ACTH) measurement, and for response to dexamethasone

	TIME	CORTISOL [μg/dL (nmol/L)]	ACTH [pg/mL (ng/L)]
Day 1	0000	12.3 (340)	—
	0800	10.2 (280)	—
Day 2	0000	15.6 (430)	—
	0800	16.7 (460)	—
Day 3	0000	16.7 (460)	—
	0800	17.0 (470)	366 (366)
Day 5	0800	14.9[a] (410)	—
Day 9	0800	13.6[b] (375)	—

[a] After 1 mg of dexamethasone
[b] After 0.5 mg of dexamethasone four times a day for 48 h

WHAT IS YOUR FINAL DIAGNOSIS?

WHICH OF THESE RESULTS ARE UNRELIABLE FOR MAKING THIS DIAGNOSIS IN PREGNANCY?

FINAL DIAGNOSIS: Cushing's syndrome caused by a glucocorticoid adenoma in the right adrenal gland

RESULTS—INTERPRETATION IN PREGNANCY

- Although serum cortisol concentrations are usually significantly higher during pregnancy due to an increase in circulating corticosteroid-binding globulin, circadian patterns are maintained. This means that the absence of diurnal variation may be used as a diagnostic test for Cushing's syndrome in pregnancy
- In this patient, the morning serum cortisol results were not elevated to the degree expected for gestational age, but the diagnostic feature was the lack of diurnal variation
- The (1-mg) overnight dexamethasone suppression test should not be used because it is often unreliable in pregnancy
- Although the low- and high-dose dexamethasone tests have been recommended, their usefulness may be limited by a change in sensitivity to dexamethasone during pregnancy
- Serum ACTH increases during pregnancy, which limits its value in the differential diagnosis of Cushing's syndrome. (ACTH measured 2 wk postpartum in this patient was undetectable)
- The corticotrophin-releasing hormone test and magnetic resonance imaging have both been reported to be useful in the differential diagnosis of Cushing's syndrome in pregnancy

KEY FEATURES

- Cushing's syndrome in pregnancy is rare, but the diagnosis is important because of the high rate of both maternal and fetal complications
- The diagnosis is difficult because many of the recommended protocols are confounded by the normal physiological hypercortisolemia of pregnancy
- Primary adrenocortical tumors rather than pituitary Cushing's disease predominate in pregnancy
- Differentiation may be complicated by the increase in serum ACTH concentration during pregnancy

POINTS TO REMEMBER

- Physiological changes affecting glucocorticoid status during pregnancy
- Appropriate biochemical tests for diagnosis
- Interpretation of these tests during pregnancy

REFERENCES

1. Guilhaume B, Sanson ML, Billaud L, Bertagna X, Laudat MH, Luton JP. Cushing's syndrome and pregnancy: aetiologies and prognosis in twenty-two patients. Eur J Med 1992;1:83–9.
2. Lockitch G. Clinical biochemistry of pregnancy. Crit Rev Clin Lab Sci 1997;34:67–139.
3. Nolten WF, Lindheimer MD, Rueckert PA, Oparil S, Ehrlich EN. Diurnal patterns and regulation of cortisol secretion in pregnancy. J Clin Endocrinol Metab 1980;51:46–72.
4. Perry LA, Grossman AB. The role of the laboratory in the diagnosis of Cushing's syndrome. Ann Clin Biochem 1997;34:349–59.
5. Ross RJ, Chew SL, Perry L, Erskine K, Medbak S, Ashfar F. Diagnosis and selective cure of Cushing's disease during pregnancy by transsphenoidal surgery. Eur J Endocrinol 1995;132:722–6.

THIS CASE WAS PRESENTED BY MRS. RUTH LAPWORTH, WILLIAM HARVEY HOSPITAL, ASHFORD, KENT, UK

4. A vomiting neonate

PRESENTATION

HISTORY OF PRESENTING COMPLAINT
- A 3-wk-old neonate presented at the accident and emergency department with a history of continual crying, forceful vomiting after all feeds, and weight loss since birth

PAST MEDICAL HISTORY
- Born at 39/40 after a normal pregnancy
- Well at birth, although noted to be very windy with some feeding difficulty and regurgitation of small amounts of clotted milk
- Bottle fed with formula milk
- Recorded as having been unwell since day 2–3, vomiting after every feed (nonprojectile, always milky, nonbilious)
- A few flecks of blood had been seen in vomit recently
- Weight loss: birth weight 8 pounds (3.64 kg); weight at presentation 7.2 pounds (3.28 kg)
- No diarrhea

FAMILY HISTORY
- First child of consanguineous (first-cousin) parents
- Mother has had two spontaneous miscarriages (at 8/40) and a termination (at 11/40 due to a tubal pregnancy)
- Infant lives with mother, father, paternal grandparents, and two uncles
- No other family history of metabolic disease or pregnancy loss

ON EXAMINATION
- Floppy, poorly responsive baby
- Dry with sunken fontanelle
- Apyrexial
- Thin and scrawny, with loose skin folds
- Not dysmorphic
- Chest clear
- Abdomen soft and nontender
- Pigmented genitalia

WHAT IS YOUR PROVISIONAL DIAGNOSIS?

WHAT INVESTIGATIONS WOULD YOU REQUEST?

PROVISIONAL DIAGNOSES
- Pyloric stenosis
- Congenital adrenal hyperplasia
- Other metabolic disturbance

INITIAL INVESTIGATIONS
Biochemistry
Blood

pH	7.42	(H⁺ 38.5 nmol/L)
Bicarbonate	19.6 mEq/L	(19.6 mmol/L)

Serum

Sodium	97 mEq/L	(97 mmol/L)
Potassium	7.7 mEq/L	(7.7 mmol/L)
Urea	31 mg/dL	(10.9 mmol/L)
Creatinine	0.8 mg/dL	(61 µmol/L)
Chloride	74 mEq/L	(74 mmol/L)
Calcium	11.7 mg/dL	(2.92 mmol/L)
Magnesium	2.29 mg/dL	(0.94 mmol/L)
Total bilirubin	1.9 mg/dL	(32 µmol/L)
Alkaline phosphatase	571 IU/L	(571 IU/L)
Alanine aminotransferase	33 IU/L	(33 IU/L)
Glucose	68 mg/dL	(3.8 mmol/L)
Lactate	16.2 mg/dL	(1.8 mmol/L)

Hematology

Hemoglobin	16.6 g/dL	(166 g/L)
Leukocyte count	$12.1 \times 10^3/\mu L$	($12.1 \times 10^9/L$)
Platelets	$559 \times 10^3/\mu L$	($559 \times 10^9/L$)

Abdominal palpation: no evidence of pyloric stenosis despite very long palpation

---

WHAT IS YOUR PROVISIONAL FINAL DIAGNOSIS?

WHAT FURTHER INVESTIGATIONS WOULD YOU REQUEST, IF ANY?

FURTHER INVESTIGATIONS

- On further examination, the endocrinologist noted that the baby had pigmented genitalia, nipples, and linea alba (line down the abdomen from sternum to pubis)

17α-Hydroxyprogesterone	>10,000 ng/dL (>300 nmol/L)
Urinary steroid profile	17α-Hydroxyprogesterone and 21-deoxycortisol metabolites were detected

FINAL DIAGNOSIS: Salt-losing congenital adrenal hyperplasia, due to 21-hydroxylase deficiency

MANAGEMENT

- Initially, 0.9% saline was administered as replacement therapy followed by 0.9% saline/10% dextrose as maintenance therapy. Hydrocortisone [12.5 mg every 6 h intravenously (i.v.)] was started. Electrolytes were checked every 2 h, and the baby was weighed every 6 h
- Potassium chloride (2.5 mmol/kg/24h) was subsequently introduced into the hydration therapy
- Once the baby's plasma sodium concentration had reached >120 mEq/L (>120 mmol/L), his maintenance fluids were replaced progressively by milk feedings. He continued on i.v. hydrocortisone (12.5 mg/6 h), and his biochemistry was monitored every 4 h
- Once the biochemistry had normalized, the baby was changed to oral hydrocortisone (5 mg, 3 times daily), oral fludrocortisone (50 µg/d), and oral sodium supplements (5 mEq/kg/d)
- The baby started to feed well and gained weight. He was no longer vomiting and remained apyrexial. He was discharged 7 d after admission on hydrocortisone, fludrocortisone, and oral sodium supplementation

KEY FEATURES

- In the classical form of 21-hydroxylase deficiency, which occurs in ~1:12,000 to ~1:13,000 births, affected females are born with virilized external genitalia due to excessive production of adrenal androgens. The degree of virilization varies, but may be sufficiently severe to result in sex misassignment. At birth, affected males usually have normal genitalia apart from increased scrotal skin pigmentation. Untreated, both males and females undergo progressive postnatal virilization
- In 60–75% of cases of classical 21-hydroxylase deficiency, salt-wasting occurs, due to an associated defect in aldosterone synthesis. This is characterized by hyponatremia, hyperkalemia, and inappropriate natriuresis
- Salt-wasting is potentially fatal and often presents as an acute crisis during the second or third week of life
- Increased plasma adrenocorticotropic hormone (ACTH) and androgen concentrations lead to pigmentation and virilization, respectively
- Analysis of the urinary steroid profile by a specialist laboratory would show elevations in intermediary metabolites consistent with the enzyme deficiency

MANAGEMENT OF CLASSICAL 21-HYDROXYLASE DEFICIENCY

- Acute treatment of an infant in a salt-losing crisis with glucocorticoid and mineralocorticoid therapy and sodium replacement is necessary
- Long-term therapy involves administration of glucocorticoids (hydrocortisone) to inhibit overproduction of androgens, driven by increased synthesis of ACTH, and thus prevent progressive virilization
- Prognosis is usually 'reasonable,' although patients are of short stature if poorly controlled

- Girls are often overweight and have polycystic ovaries
- Girls with congenital adrenal hyperplasia also have a high incidence of psychosexual problems

POINTS TO REMEMBER
- Appreciate the basic metabolic pathways of adrenal steroid hormone synthesis and the causes of congenital adrenal hyperplasia
- Remember, in 60–75% cases of classical 21-hydroxylase deficiency, salt-wasting occurs and can be life threatening
- Salt-losing congenital adrenal hyperplasia should be suspected in an infant who fails to thrive, especially in females with ambiguous genitalia
- 17α-hydroxyprogesterone (17OHP) is useful in the diagnosis of classical 21-hydroxylase deficiency, but specimen collection should be delayed until at least the 3rd day because 17OHP is normally high during the first 2–3 days of life

REFERENCES
1. Levine LS, DiGeorge AM. Disorders of the adrenal glands. In: Behrman RE, Kliegman RM, Jenson HB, eds. Nelson textbook of pediatrics, 16th ed. Philadelphia: WB Saunders, 2000:1722–44.
2. Drury PL. Adrenal disorders. In: Marshall WJ, Bangert SK, eds. Clinical biochemistry—metabolic and clinical aspects. New York: Churchill Livingstone, 1995:315–29.

THIS CASE WAS PRESENTED BY DR. MICHELLE BIGNELL AND DR. JEREMY KIRK, BIRMINGHAM CHILDREN'S HOSPITAL, BIRMINGHAM, UK

5. A well-developed boy

PRESENTATION

HISTORY OF PRESENTING COMPLAINT
- A 6-y-old boy was referred to a pediatric endocrinologist with a suspected growth disorder because he was substantially taller than the other pupils in his class

DIRECT QUESTIONING
- Good health

PAST MEDICAL HISTORY
- Normal delivery at term
- Divergent squint

SOCIAL HISTORY
- Performing reasonably well at school

FAMILY HISTORY
- No family history of note
- Both parents of average height

MEDICATION
- None

ON EXAMINATION
- Height >99th centile for age
- Weight at 50th centile
- Penile length and volume increased (Tanner stage 2), but testes normal for age (2–3 mL)
- Normotensive
- No other abnormal findings

WHAT ARE YOUR PROVISIONAL DIAGNOSES?

WHAT INVESTIGATIONS ARE INDICATED IN THIS PATIENT?

PROVISIONAL DIAGNOSES
- Congenital adrenal hyperplasia
- Virilizing neoplasm

INITIAL INVESTIGATIONS

Dipstick urinalysis
Glucose	Negative
Protein	Negative
Blood	Negative

Serum
Sodium	137 mEq/L	(137 mmol/L)
Potassium	3.7 mEq/L	(3.7 mmol/L)
Urea	19 mg/dL	(6.7 mmol/L)
Creatinine	0.53 mg/dL	(47 µmol/L)
Testosterone	80 ng/dL	(2.8 nmol/L)
Aldosterone	280 pg/mL	(771 pmol/L)
Cortisol (random)	1.5 µg/dL	(41 nmol/L)

Plasma
Renin activity	18.6 ng/h/mL	(22.7 pmol /h/mL)

WHAT FURTHER INVESTIGATIONS, IF ANY, WOULD YOU REQUEST?

HAVE YOU MADE ANY CHANGES TO YOUR PROVISIONAL DIAGNOSIS?

WHAT IS YOUR DIAGNOSIS?

WORKING DIAGNOSIS
- Non-salt-losing congenital adrenal hyperplasia
- Subsequently he went on to have an adrenocorticotropic hormone (ACTH) stimulation test:

	CORTISOL [μg/dL (nmol/L)]	17-OHP [ng/dL (nmol/L)]	ANDROSTENEDIONE [ng/dL (nmol/L)]
0 min	0.87 (24)	>1000 (>300)	378 (13.2)
30 min	1.27 (35)		
60 min		>1000 (>300)	461 (16.1)

17-OHP, 17α-hydroxyprogesterone

FINAL DIAGNOSIS: Congenital adrenal hyperplasia due to 21-hydroxylase deficiency (nonclassical)

MANAGEMENT
- He was started on oral replacement hydrocortisone
- Fludrocortisone was added because of the raised plasma renin activity

KEY FEATURES

PRESENTATION
- Ninety percent of patients with congenital adrenal hyperplasia have 21-hydroxylase deficiency
- The classical presentations are virilization or ambiguous genitalia in female neonates and salt-losing crisis in the first few days of life
- Nonclassical presentations occur later in the first decade due to virilization of females and precocious puberty in males. They do not have salt-losing crises

BIOCHEMISTRY (see figure)
- 21-Hydroxylase deficiency blocks the synthesis of cortisol and aldosterone
- Lack of negative feedback by cortisol leads to increased ACTH concentrations and high concentrations of the steroid precursors, which feed into the sex steroid pathway and also have some androgenic activity themselves
- This results in virilization of females and precocious puberty in males
- Lack of aldosterone leads to renal sodium and water loss, which, if unrecognized, results in circulatory collapse and death
- Other enzyme deficiencies can occur, such as 11-β-hydroxylase and 3-β-hydroxysteroid dehydrogenase. The clinical features of these disorders vary

DIAGNOSIS
- Elevated concentrations of 17-hydroxyprogesterone before and 60 min after ACTH stimulation are diagnostic for 21-hydroxylase deficiency
- High plasma renin activity indicates a requirement for fludrocortisone replacement even though, as in this case, sufficient aldosterone is being produced to prevent hyponatremia

FIGURE   METABOLIC PATHWAY SHOWING THE PRODUCTION OF SEVERAL STEROID HORMONES FROM CHOLESTEROL AND THE ROLE OF KEY ENZYMES
DHEA, dehydroepiandrosterone

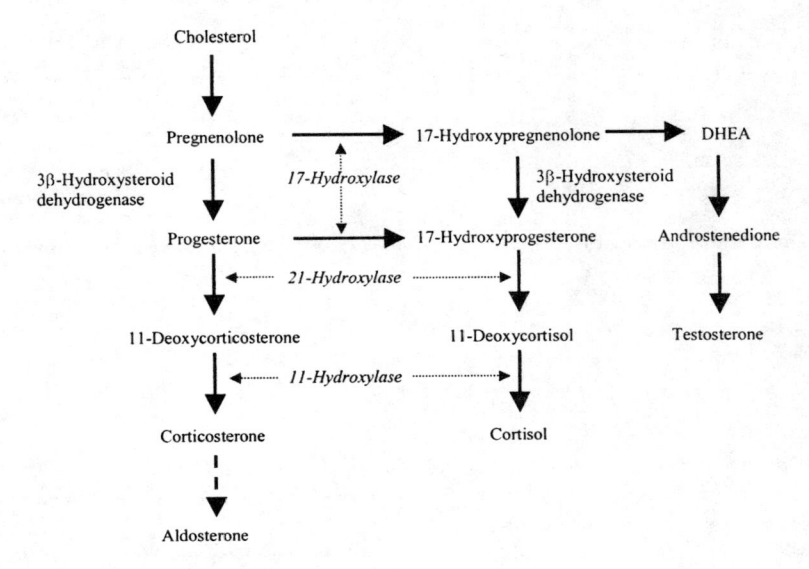

```
                        Cholesterol
                             │
                             ▼
                       Pregnenolone ──────► 17-Hydroxypregnenolone ──────► DHEA
  3β-Hydroxysteroid          │  17-Hydroxylase    │  3β-Hydroxysteroid       │
  dehydrogenase              ▼                    ▼  dehydrogenase           ▼
                       Progesterone ──────► 17-Hydroxyprogesterone     Androstenedione
                             │  21-Hydroxylase    │                         │
                             ▼                    ▼                         ▼
                  11-Deoxycorticosterone      11-Deoxycortisol          Testosterone
                             │  11-Hydroxylase    │
                             ▼                    ▼
                       Corticosterone          Cortisol
                             │
                             ▼
                        Aldosterone
```

POINTS TO REMEMBER
- Be aware of the causes of precocious puberty
- Understand the causes of congenital adrenal hyperplasia
- Understand how to use laboratory investigations to diagnose the cause of congenital adrenal hyperplasia

REFERENCES
1. Levine LS, DiGeorge AM. Adrenal disorders and genital abnormalities. In: Behrman RE, Kliegman RM, Jenson HB, eds. Nelson textbook of pediatrics, 16th ed. Philadelphia: WB Saunders, 2000:1729–37.
2. Speiser PW, New MI. An update of congenital adrenal hyperplasia. In: Lifshitz F, ed. Pediatric endocrinology, a clinical guide, 2nd ed. New York: Marcel Dekker, 1990:307–31.

THIS CASE WAS PRESENTED BY DR. STEVEN C. MARTIN, WEST SUFFOLK HOSPITAL, BURY ST EDMUNDS, UK

6. A gastrointestinal starting point

PRESENTATION

HISTORY OF PRESENTING COMPLAINT
- A 40-y-old woman presented with acute onset of diarrhea and vomiting of 1-d duration

DIRECT QUESTIONING
- She mentioned that her 1-y-old son and husband also had diarrhea 2 d before her symptoms
- She confirmed that she had vomited ~20 times over a period of 8 h. Vomit and stool did not show any signs of blood
- She also complained of episodes of extreme tiredness and fatigue

PAST MEDICAL HISTORY
- 15-y history of atopic eczema and mild asthma
- Removal of benign lump from her right breast
- Normal pregnancy a year before this admission

ON EXAMINATION
- She looked extremely unwell and was pyrexic with a temperature of 38 °C
- She was profoundly dehydrated and hypotensive with blood pressure of 65/40 mmHg
- She was admitted to the intensive-care unit under the care of the gastroenterology team
- She was put on fluid replacement and required a stat dose of dopamine

---

WHAT IS YOUR PROVISIONAL DIAGNOSIS OF HER PRESENT COMPLAINT?

WHAT INVESTIGATIONS WOULD YOU REQUEST?

PROVISIONAL DIAGNOSIS
- Sepsis
- Gastroenteritis (viral)
- Cholecystitis
- Pancreatitis

INITIAL INVESTIGATIONS
On admission
Serum

Sodium	134 mEq/L	(134 mmol/L)
Potassium	4.3 mEq/L	(4.3 mmol/L)
Chloride	115 mEq/L	(115 mmol/L)
Bicarbonate	9 mEq/L	(9 mmol/L)
Urea	20 mg/dL	(7.2 mmol/L)
Creatinine	0.9 mg/dL	(84 µmol/L)
C-reactive protein	6.9 mg/dL	(69 mg/L)
Albumin	4.8 g/dL	(48 g/L)
Amylase	158 IU/L	(158 IU/L)
Creatine kinase	46 IU/L	(46 IU/L)
Glucose	148 mg/dL	(8.2 mmol/L)
Hemoglobin	17.7 g/dL	(177 g/L)

- Blood gas analysis with above results confirmed that she had compensated metabolic acidosis
- Patient remained hypotensive despite several liters of fluid replacement

Biochemistry on day 3
Serum

Sodium	117 mEq/L	(117 mmol/L)
Potassium	5.6 mEq/L	(5.6 mmol/L)
Chloride	91 mEq/L	(91 mmol/L)
Bicarbonate	17 mEq/L	(17 mmol/L)
Urea	9 mg/dL	(3.2 mmol/L)
Creatinine	0.78 mg/dL	(70 µmol/L)
Creatine kinase	219 IU/L	(219 IU/L)

Urine

Sodium	107 mEq/L	(107 mmol/L)
Potassium	22 mEq/L	(22 mmol/L)

- Stool examination was negative
- Ultrasound of liver, gallbladder, pancreas, and spleen showed no focal abnormality
- Barium follow through: No abnormality seen. Limited view of the stomach and duodenum revealed no abnormality

---

WHAT FURTHER INVESTIGATIONS, IF ANY WOULD YOU REQUEST?

HAVE YOU MADE ANY CHANGES TO YOUR PROVISIONAL DIAGNOSIS?

WHAT IS YOUR DIAGNOSIS?

WORKING DIAGNOSES
- Metabolic or endocrine causes of hypotension
- Adrenal insufficiency due to long-term topical steroid use
- Addisonian crisis due to Addison's disease

FURTHER INVESTIGATIONS
- Pituitary adrenal axis
- Other pituitary hormones including thyrotropin (TSH), follicle-stimulating hormone (FSH), luteinizing hormone (LH), lactogenic hormone (LGH), prolactin
- Estradiol, free thyroxine (T_4)
- Organ-specific antibody screen
- Short Synacthen (cosyntropin) test

Serum
0900 cortisol	7.9 µg/dL	(219 mmol/L)
Adrenocorticotropic hormone (ACTH)	136 pg/mL	(136 ng/L)
Free T_4	2.2 ng/dL	(9.3 pmol/L)
TSH	>50 µU/mL	(>50 mU/L)
FSH	6.5 mIU/mL	(6.5 IU/L)
LH	3.8 mIU/mL	(3.8 IU/L)
Estradiol	69 pg/mL	(254 pmol/L)
Prolactin	30 ng/mL	(30 µg/L)
Antibodies to adrenal cortex	Positive	
Antibodies to thyroid peroxidase	Positive	

Short Synacthen test

TIME	CORTISOL	
0 min	5.7 µg/dL	(158 nmol/L)
30 min	6.3 µg/dL	(175 nmol/L)
60 min	5.9 µg/dL	(165 nmol/L)

FINAL DIAGNOSIS: Polyglandular autoimmune syndrome type II with Addison's disease and autoimmune hypothyroidism

MANAGEMENT
- Fluid replacement, initially with stat dose of dopamine to restore normal volume and to correct hypotension
- Hydrocortisone replacement three times a day (dose at 0700, 10 mg; at 1300, 10 mg; at 1800, 5 mg)
- Fludrocortisone 150 µg once daily
- The patient was commenced on T_4 treatment (10 µg/d, increasing to 50 µg/d at the time of discharge)

KEY FEATURES
PRESENTATION
- Addison's disease is a relatively rare condition, with an incidence in the developed world of 0.8 cases per 100,000 population
- Diagnosis of Addison's disease is often only made when the patient presents with a critical illness. Sometimes diagnosis is made postmortem. This delay is because recognition of cortisol deficiency is difficult because symptoms and signs are nonspecific, such as tiredness, weakness, and spells of dizziness
- Cortisol rather than aldosterone deficiency is responsible for most of the clinical symptoms
- Congenital adrenal hyperplasia and hypoplasia are not normally included as causes of Addison's disease
- At present, the major cause of primary adrenocortical failure is autoimmune adrenalitis (accounting for 70% of all cases). Adrenocortical failure can occur as part of an acquired immune deficiency syndrome, and also the incidence of tuberculosis is increasing. As a consequence the prevalence and etiology of Addison's disease may change
- Several other diseases are associated with autoimmune Addison's disease and may thus suggest the diagnosis, such as hypothyroidism, hypoparathyroidism, and diabetes mellitus
- Organ-specific autoimmune disorders are classified as autoimmune polyglandular deficiency type I and type II (see table)
- Human leucocyte antigens (HLA) DR3 and DR4 are strongly associated with autoimmune Addison's disease type II
- Presentation of adrenocortical failure is usually marked by hyponatremia, hyperkalemia, and a raised urea concentration. However, hyperkalemia is less frequent if there is no marked aldosterone deficiency. At presentation this patient's sodium, potassium, and urea concentrations did not match the degree of hypotension. This may reflect the loss of electrolyte through vomiting and diarrhea
- This patient's 0900 cortisol was in the lower limit of the reference range, and her ACTH was elevated. These results confirmed that she had adrenocortical insufficiency, but because she had been using topical steroid for her eczema and inhaling steroid for asthma, her adrenal failure could have been due to prolonged use of the exogenous steroid. Differential diagnosis for the etiology of Addison's disease should consider this as one of the causes of adrenocortical failure. For this patient, the differential diagnosis did not pose a problem because she demonstrated antibodies to adrenal cortex and thyroid peroxidase, thus making the diagnosis of autoimmune polyglandular deficiency type II possible without performing a depot Synacthen stimulation test
- Her fasting glucose concentration was in the reference range and her gonadotrophins and estradiol concentrations were within the reference range, confirming that her ovaries and pancreas were intact
- Hyperprolactinemia in adrenocortical deficiency is an incidental finding. Prolactin concentration can be as high as 100 ng/mL (100 µg/L)

TABLE
Classification of autoimmune polyglandular deficiency syndrome

AUTOIMMUNE POLYGLANDULAR DEFICIENCY TYPE I	AUTOIMMUNE POLYGLANDULAR DEFICIENCY TYPE II
Addison's disease	Addison's disease
Chronic mucocutaneous candidiasis	Primary hypothyroidism
Hypoparathyroidism	Primary hypogonadism
	Type 1 diabetes

BIOCHEMISTRY
- Addison's disease results from primary adrenocortical failure. Cortisol deficiency results in uncontrolled excess secretion of ACTH and melanocyte-stimulating hormone (MSH)
- MSH is responsible for the pigmentation of the skin found in Addison's disease
- Confirmation of the diagnosis of Addison's disease is normally made by measuring early morning concentrations of cortisol and ACTH. Cortisol may be low or within the reference range, whereas ACTH concentrations are always high. Raised ACTH concentrations also aid in the differential diagnosis of primary and secondary adrenocortical insufficiency
- Confirmation is also aided by determining plasma cortisol responsiveness to Synacthen. In normal patients, after intramuscular administration of 250 μg of Synacthen, cortisol concentrations should increase >7.2 μg/dL (>200 nmol/L) at 30 or 60 min
- For the differential diagnosis of primary adrenal failure and secondary causes of adrenal failure, the depot Synacthen test can be helpful
- Serum aldosterone concentration may be normal, but renin is elevated. Patients with Addison's disease have an increase basal value of vasopressin and a decreased osmotic threshold
- Hyperprolactinemia in adrenocortical deficiency is an incidental finding. Prolactin concentrations can be as high as 100 ng/mL (100 μg/L)

DIAGNOSIS
Features with high diagnostic significance include:
- Hyponatremia and hyperkalemia: Hyponatremia is present in ≥90% of severely cortisol-deficient patients; urinary sodium excretion is high. Hyperkalemia is less frequent than hyponatremia. Hyperkalemia is generally present if there is aldosterone deficiency as well
- Postural hypotension: Although a fall in blood volume in adrenal insufficiency is modest, several other mechanisms contribute to the fall in blood pressure
- Pigmentation is present in >70% of cases, this results from increased melanin production in the skin and mucous membrane. Patients presenting with acute adrenal failure occurring de novo (as with septicemia) will not be pigmented
- The other invariable symptoms will be weakness, malaise, nausea (often with vomiting), and nonspecific vague epigastric abdominal pain with either constipation or diarrhea

TREATMENT
- Patients in Addisonian crisis should be treated with intravenous fluids and glucose together with glucocorticoid replacement
- Patients with chronic Addison's disease require glucocorticoid replacement and the majority also need mineralocorticoid treatment. Hydrocortisone is now regarded as the drug of choice. The dose is divided into three doses over a 24-h period and given to mimic endogenous cortisol production

POINTS TO REMEMBER

- Hyponatremia is common in a hospital population. Although the incidence of Addison's disease is low, the possibility should be explored in all patients with acute illness with no apparent cause
- Because use of topical steroids and oral steroids is common, secondary causes of adrenal insufficiency should be included in the differential diagnosis of low cortisol

REFERENCES

1. Burke CW. Primary adrenocortical failure. In: Grossman A, ed. Clinical endocrinology, 1st ed. Oxford: Blackwell, 1992:393–404.
2. Edwards CRW. Addison's disease. In: Besser GM, Thorner OM, eds. Clinical endocrinology, 2nd ed. London: Times Mirror International, 1994:9.1–9.8.
3. Neufeld M, Maclaren NK, Blizzard RM. Two types of autoimmune Addison's disease associated with different polyglandular autoimmune syndromes. Medicine 1981;60:355–62.
4. Maclaren NK, Riley WJ. Inherited susceptibility to autoimmune Addison's disease is linked to human leucocyte antigen-DR3 and/or DR4; except when associated with type 1 autoimmune polyglandular syndrome. J Clin Endocrinol Metab 1986;62:455–9.
5. Dluhy RG. The growing spectrum of HIV-related endocrine abnormalities. J Clin Endocrinol Metab 1990;70:563–5.

THIS CASE WAS PRESENTED BY MRS. KATE NOONAN, THE ROYAL LONDON HOSPITAL, BARTS AND THE LONDON NHS TRUST, LONDON, UK

7. Tired and weepy

PRESENTATION

HISTORY OF PRESENTING COMPLAINT
- A 30-y-old woman attended her primary care physician and complained of shortness of breath at rest, intense fatigue, weepiness, and early morning vomiting. She had lost 21 pounds (9.5 kg) and weighed 136 pounds (62 kg)

PAST MEDICAL HISTORY
- Two years previously she had been investigated for infertility of 2 y duration. She had a history of one miscarriage at 7 wk. No hormonal abnormalities were found on routine screening. Clomiphene was given for six menstrual cycles over 12 mo. No pregnancy was achieved
- She had been admitted through the emergency department 7 mo previously with chest pain, fever, weakness, sweating, and cough. A heart rate of 120 beats per minute was recorded. Pulmonary embolus or severe chest infection was suspected
- At that time urea and electrolytes, liver function tests, hemoglobin, and leukocyte count were all normal. She was treated with antibiotics and discharged
- Three months ago she had undergone her first attempt at in vitro fertilization. It was not successful

ON EXAMINATION
- She looked ill
- Her skin was grayish-brown
- Her skin was warm and clammy
- Her heart rate was 128 beats per minute, but regular
- She weighed 136 pounds (62 kg)
- Her abdomen was soft with no localized tenderness
- She was afebrile

WHAT IS YOUR PROVISIONAL DIAGNOSIS?

WHAT INVESTIGATIONS WOULD YOU REQUEST?

PROVISONAL DIAGNOSES
- Diabetes
- Thyrotoxicosis
- Pregnancy with hyperemesis
- Tuberculosis
- Severe viral infection
- Anemia

INITIAL INVESTIGATIONS

Blood glucose	103 mg/dL	(5.7 mmol/L)
Hemoglobin	8.0 g/dL	(80 g/L)
Mean erythrocyte volume	103 fL	
Leukocyte count	$5 \times 10^3/mm^3$ with neutropenia	(5×10^9/L)
Thyrotropin	<0.02 µU/mL	(<0.02 nU/L)
Free thyroxine	4.5 mg/dL	(58 pmol/L)
Free triiodothyronine	1558 pg/dL	(24.0 pmol/L)
Blood film	Macrocytosis and anisocytosis	
Thyroid microsomal antibodies	Strongly positive	
Serum iron	196 µg/dL	(35 µmol/L)
Vitamin B_{12}	<50 pg/mL	(< 37 pmol/L)
Serum folate	7.1 ng/mL	(16 nmol/L)
Antibodies to gastric parietal cells and intrinsic factor	Positive	

WHAT FURTHER INVESTIGATIONS, IF ANY, WOULD YOU REQUEST?

HAVE YOU MADE ANY CHANGES TO YOUR PROVISIONAL DIAGNOSIS?

WHAT IS YOUR DIAGNOSIS?

FINAL DIAGNOSIS
- Autoimmune vitamin B_{12} deficiency
- Pernicious anemia (PA)
- Thyrotoxicosis

MANAGEMENT
- She was given a course of vitamin B_{12} injections
- Her primary-care physician started her on carbimazole. This was changed to propylthiouracil (PTU) because she was hoping for a pregnancy
- Within 1 wk, she was considerably improved. After 6 wk on treatment, she was euthyroid and her hemoglobin concentration had returned to normal
- She returned for further in vitro fertilization 6 mo later with ovulation induction and subsequent embryo implantation. Her pregnancy progressed normally and she delivered a normal healthy boy at term

KEY FEATURES

PRESENTATION
- She presented with breathlessness, fatigue, and tachycardia—common symptoms of tissue anoxia, which occurs in anemia and thyrotoxicosis
- PA can present with anemia, diarrhea, and malabsorption, paresthesia and numbness due to peripheral neuropathy, and personality changes

EPIDEMIOLOGY
- PA is the most common cause of vitamin B_{12} deficiency. It is more common in women than in men and 20% of cases occur among relatives. PA may be associated with autoimmune endocrinopathies including Graves disease and hyperthyroidism

BIOCHEMISTRY
- Dietary requirements of Vitamin B_{12} are 2 µg/d. The chief source is animal products, and none is present in leafy vegetables. Vitamin B_{12} is water soluble and is a complex of cobalamin and four substituted pyrrole rings. Intrinsic factor is produced by gastric parietal cells; it is a 60-kDa glycoprotein that binds avidly to vitamin B_{12} and is necessary for its absorption via protein receptors in the terminal ileum

FIGURE   ABSORPTION OF VITAMIN B$_{12}$

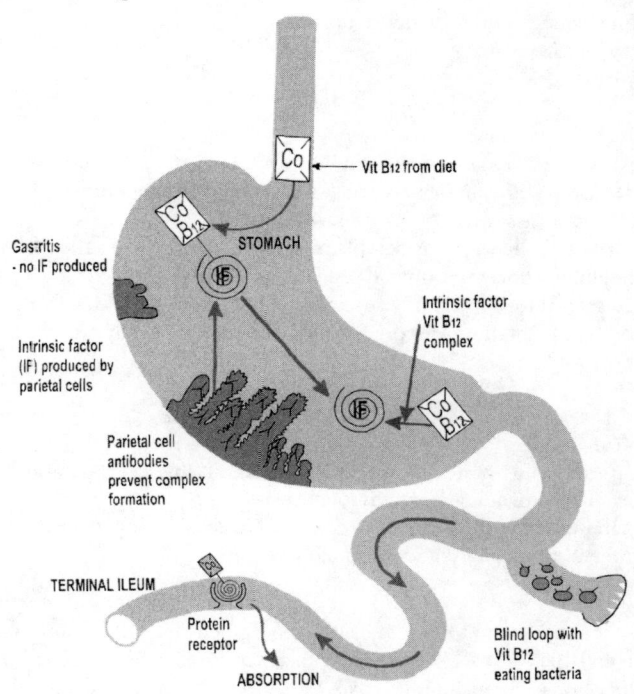

Vit B12 from diet
STOMACH
Gastritis - no IF produced
Intrinsic factor (IF) produced by parietal cells
Intrinsic factor Vit B12 complex
Parietal cell antibodies prevent complex formation
TERMINAL ILEUM
Protein receptor
ABSORPTION
Blind loop with Vit B12 eating bacteria

- Four years previously after the miscarriage macrocytosis had been reported on her blood film. It usually takes 4 y to develop anemia after withdrawal of the vitamin. Vitamin B$_{12}$ functions as a coenzyme and is essential for the methylation of homocysteine to methionine (an essential amino acid), which in turn ensures active DNA synthesis. Patients with Vitamin B$_{12}$ deficiency show macrocytosis and neutropenia and thrombocytopenia in the blood film
- Macrocytosis is visible on the blood film, but all cells in the body are affected by vitamin B$_{12}$ deficiency. Cells most affected are those with a rapid turnover such as bone marrow and those in the gastrointestinal tract. It is interesting and possibly significant that only 14 eggs were harvested from this woman during her first assisted conception and that 12 of these survived <24 h
- Intrinsic factor antibodies are found in the serum of 90% of patients with pernicious anemia. Their presence makes further absorption studies unnecessary. They occasionally occur in the serum of patients with other autoimmune disorders such as thyroid disorders, type 1 diabetes mellitus, and myasthenia gravis
- Forty-eight percent of patients with PA have thyroid disease. This may be because both are autoimmune disorders, and thyroid and stomach are both derived embryologically from the foregut
- Thyrotoxicosis is also associated with irregular menstruation and infertility. Two years previously, at initial investigations for infertility, the thyrotropin concentration had been normal at 0.87 µU/mL (0.87 mIU/L)

MANAGEMENT

- Vitamin B$_{12}$ 1000 µg is given by intramuscular injection to correct the anemia and at regular 3-mo intervals thereafter
- Medical treatment of hyperthyroidism with antithyroid drugs is usually undertaken over a 12- to 18-mo period. The two drugs available, carbimazole and PTU, both cross the placenta and may cause neonatal goiter and hypothyroidism. Thyroid development takes place during the first 10 wk of fetal life
- Carbimazole has also been associated with aplasia cutis in the neonate (a teratogenic effect) and is not recommended for use in pregnancy. Radioiodine should not knowingly be given in pregnancy, and thus patients who are pregnant or trying to become pregnant are treated with PTU
- The need for antithyroid treatment often declines as the pregnancy progresses. It may be discontinued in the third trimester

POINTS TO REMEMBER

- Pernicious anemia is an autoimmune disorder, and 48% of patients have thyroid disease
- There is documented association between pernicious anemia and infertility and between hyperthyroidism and infertility
- A low hemoglobin and hyperthyroidism both lead to tissue anoxia with breathlessness, tachycardia, irritability, lassitude, and weakness
- Vitamin B$_{12}$ is essential as a co-enzyme for methylation of homocysteine to methionine, which ensures active DNA synthesis for cell maturation
- Treatment of PA is usually by intramuscular injection at regular intervals and continues lifelong
- PTU is the drug of choice in hyperthyroidism in pregnancy

REFERENCES

1. Burns RW, Burns TW. Pancytopenia due to vitamin B12 deficiency associated with Graves disease. Mol Med 1996;93(7):368–72.
2. Campbell BA. Megaloblastic anaemia in pregnancy. Clin Obstet Gynaecol 1995;38:455–62.
3. Menachem Y, Cohen AM, Mittleman M. Cobalamin deficiency and infertility [Letter]. Am J Hematol 1994;46:152.
4. O'Doherty MJ, McElhatton PR, Thomas SHL. Treating thyrotoxicosis in pregnant or potentially pregnant women [Editorial]. Br M J 1999;318:5–6.
5. Toh BH, van Driel IR, Gleeson PA. Mechanisms of disease: pernicious anaemia. New Engl J Med 1997;337;20:1441–8.

THIS CASE WAS PRESENTED BY DR. HAZEL WILKINSON, YORK DISTRICT HOSPITAL, YORK, UK

8. A surprising testosterone value

PRESENTATION

HISTORY OF PRESENTING COMPLAINT
- A 42-y-old woman weighing 145 pounds (66.5 kg), and complaining of hirsutism, was referred from her primary care physician to a consultant gynecologist
- The condition had become worse in the last 18 mo
- Menstruation was occurring about four times per year
- The provisional diagnosis by the primary care physician was polycystic ovarian disease (PCOD)

PAST MEDICAL HISTORY
- She first noticed hirsutism as a teenager
- Menstruation had rarely been more frequent than every other month
- In her 20s, she was referred for treatment of infertility
- She remains childless

INITIAL INVESTIGATIONS
- A blood sample was sent to the laboratory, and the results obtained were as follows

Serum
Luteinizing hormone (LH)	15 mU/mL	(15 U/L)
Follicle-stimulating hormone (FSH)	8.2 mU/mL	(8.2 U/L)
Testosterone	1037 ng/dL	(36 nmol/L)
Estradiol 17β	27 pg/mL	(100 pmol/L)
Sex hormone–binding globulin (SHBG)	1.36 mg/L	(23 nmol/L)
Free androgen index (FAI)	763	(156)
(Calculated from testosterone and SHBG)		

TO WHAT EXTENT ARE THESE RESULTS COMPATIBLE WITH THE GP'S PROVISIONAL DIAGNOSIS?

WHAT OTHER DIAGNOSES OR EXPLANATIONS SHOULD BE CONSIDERED TO ACCOUNT FOR THESE RESULTS? EVALUATE EACH POSSIBILITY FROM CURRENT KNOWLEDGE OF THE CASE.

WHAT FURTHER INVESTIGATIONS WOULD YOU REQUEST?

COMPATIBILITY WITH PRIMARY-CARE PHYSICIAN'S PROVISIONAL DIAGNOSIS
- The clinical history of infrequent menses, hirsutism, and infertility over a long period of time is typical of PCOD
- The pattern of a raised LH in the presence of a normal FSH (a high LH-FSH ratio, often >3) is frequently found in patients with PCOD
- Raised androgens and FAI are expected findings in known PCOD cases. This also gives rise to a suppressed SHBG, as in this case
- However, the very high testosterone raises serious questions about the validity of this diagnosis. The result is at the upper end of the male range. It is far higher than in classical PCOD

ADDITIONAL INVESTIGATIONS
BIOCHEMICAL
- The following biochemical investigations were suggested in order to understand the nature and possible source of circulating androgens:

Plasma cortisol	13.6 µg/dL	(375 nmol/L)
Serum dehydroepiandrosterone sulphate	185 µg/mL	(5.0 µmol/L)
Serum androstenedione	169 ng/dL	(5.9 nmol/L)

OTHER
Computed tomography scans of adrenals and ovary
- Both ovaries appeared normal in size and content. No evidence of PCOD
- No evidence of adrenal mass, lesion, or hyperplasia

FURTHER INFORMATION
- Laparoscopy was performed, and the following noted at the time of the procedure
  - Normal-looking vulva with hypertrophied clitoris. Normal vagina and cervix
  - Yellowish area on right ovary (possibly suggestive of fibroma)
- Both ovaries biopsied and samples sent for histology with the following results
  - Left ovary—no follicles or stroma included in biopsy sample; origin from a cortical fibroma cannot be excluded; there is no evidence of malignancy
  - Right ovary—normal ovarian cortical stroma covered with serous epithelium; no stromal luteinization is seen; no follicles are included; no evidence of neoplasia
- Because no abnormality had been found that would easily account for such a high testosterone, further explanations were sought. The two favored by clinical colleagues were congenital adrenal hyperplasia and androgen insensitivity (testicular feminization—that is, patient was a genetic male)
- Testicular feminization was referred to in a letter to the patient's primary care physician with the comment that chromosome karyotyping would be undertaken

---

IS CONGENITAL ADRENAL HYPERPLASIA OR TESTICULAR FEMINIZATION SYNDROME A FEASIBLE DIAGNOSIS?

WHAT IS YOUR DIAGNOSIS?

WHAT ADVICE WILL YOU GIVE TO YOUR CLINICAL COLLEAGUE?

DIAGNOSES TO RULE OUT

CONGENITAL ADRENAL HYPERPLASIA

- The enzyme deficiencies that compromise the production of cortisol (21- and 11-hydroxylase deficiencies are the most common) cause accumulation of androgens. These fully functioning androgen metabolic pathways increase in prominence under continued adrenal stimulation in the absence of appropriate ACTH control by cortisol. Although androgen concentrations are increased in congenital adrenal hyperplasia, a testosterone result of the magnitude found in this patient would not be expected. 17α-Hydroxyprogesterone measurement did not support the diagnosis [133 ng/dL (4.0 nmol/L)]

TESTICULAR FEMINIZATION SYNDROME

- Individuals with this syndrome are genetically male but lack the means of expressing the androgen activity. Because this lack includes the in utero androgen influence that governs the differentiation of male from female reproductive organs, affected individuals appear phenotypically female. They are, however, infertile, lacking functional gonads. The absence of androgen expression is caused by failure in the cell receptor response to androgens. There is also a spectrum of conditions in which varying degrees of androgen insensitivity are expressed. These arise from point mutations in genes coding for parts of the androgen receptor mechanism. The expression of androgen activity is via the testosterone metabolite 5α-dihydrotestosterone under the influence of the enzyme testosterone 5α-reductase. This 'testosterone activation' occurs in peripheral tissues near the point of action. Therefore, lack of 5α-reductase also causes androgen insensitivity, but it does not usually affect sexual differentiation, presumably because the receptors are intact and can respond even to very low concentrations of 5α-dihydrotestosterone or to other androgens. The gonads that arise in the classical form of testicular feminization carry a risk of developing an adenoma. They are therefore usually removed, but if early diagnosis is missed, the presentation can occur later in life with symptoms arising from the adenoma. In the present case, the facts that the patient is hirsute and also has episodes of menstruation virtually rule out this diagnosis. Nevertheless, clinical concern about missing any possibility drove the treating clinician to request chromosome analysis. She had a normal female karyotype with no mosaicism

DIAGNOSES AND EXPLANATIONS TO CONSIDER

- The sample analyzed was not from the patient in question but had become crossed with a sample from a male subject. This hypothesis is clearly a possibility, which can be checked by blood grouping the sample and comparing it with the group of the patient (if known), or by obtaining a repeat sample from the patient. Obtaining a fresh sample is the more secure method, and in this case the repeat sample gave results that were essentially the same
- Exogenous administration of large doses of testosterone. Use of androgens is well known in bodybuilders and by sports participants (in contravention of sport guidelines). Testosterone implants are increasingly being used alongside hormone replacement therapy to add maintenance of libido to the other benefits of hormone replacement therapy. Testosterone concentrations up to 288 ng/dL (10 nmol/L) have been found in women taking this form of hormone replacement therapy. In this case it would require the subject to have taken large doses of testosterone or similar compound over a long period of time. History taking did not support this hypothesis
- Testosterone-secreting tumor. Testosterone concentrations >200–230 ng/dL (7–8 nmol/L) in a female should always be assumed to indicate the presence of a tumor until proved otherwise. The two sources to consider are the ovary and the adrenal glands. In both cases, other androgens on the metabolic pathway are also raised and, because the pattern is different between the sources, the measurement of androstenedione and dehydroepiandrosterone sulphate (DHAS) helps to

distinguish the source. In ovarian tumors androstenedione is the predominant androgen and DHAS is associated with adrenal tumors. Nevertheless, only rarely do cases of adrenal or ovarian androgen-secreting tumors give rise to such high testosterone values without rapid onset of severe virilizing features

PROVISIONAL DIAGNOSES
- Exogenous androgens
  - Exogenously administered androgens are still a possible cause. For this to be the case, the patient would have had to be deliberately supplying misleading information about her lifestyle. To achieve the concentration found, it is hard to imagine exactly what she could have taken and how she could get a regular supply of pure testosterone (for example, testosterone propionate). Although other androgens may produce a response in an immunoassay for testosterone, few if any assays respond on a mole for mole basis to analogs. The concentration of analog would need to be several times the concentration of testosterone measured. It is therefore very unlikely that this is the cause
- Ovarian tumor
  - Despite failure to find specific evidence for a tumor, this diagnosis remains the most likely because:
    - Such a high testosterone, once clearly established as belonging to the patient, is unlikely to arise from any other cause
    - The fact that neither imaging procedures nor biopsy of the ovaries found a tumor is not proof that no tumor exists. A small tumor may not be seen on scan, and a biopsy could easily miss the affected area

CLINICAL ADVICE
- Reassure that the only realistic diagnosis is a tumor, most probably ovarian. It is likely to be a tumor secreting almost exclusively testosterone because other androgen metabolites were not raised
- This diagnosis was supported by the histology consultant, who was so sure of it that he recommended bilateral oophorectomy

BILATERAL OOPHORECTOMY
MORPHOLOGY OF THE REPRODUCTIVE ORGANS
- Normal Fallopian tubes
- Normal ovarian structure and appearance
- Left ovary slightly enlarged; small, firm lesion on the surface

HISTOLOGY
- Right ovary: Mild cortical stromal hyperplasia. No evidence of malignancy
- Left ovary: Leydig cell tumor and cortical fibroma

LEYDIG CELL TUMORS
- Steroid cell tumors represent 0.1% of ovarian tumors
- Leydig cell tumors represent 20% of steroid cell tumors
- About 80% of Leydig cell tumors show androgenic manifestations
- In some patients the tumor is present for many years before diagnosis
- Androgenic changes are typically less abrupt in onset and milder than tumors involving Sertoli cells

- Leydig cell tumors usually give rise to a high testosterone concentration
- Tumors are characteristically small and can be hard to identify by imaging
- These tumors are rarely metastatic

DISCUSSION

- This patient had longstanding hirsutism, getting worse only in the last 18 mo. Considering the high value of testosterone found, it seems remarkable that more signs of virilism were not seen. Thus, we are considering a slowly developing condition in which the body appears to adapt somewhat to the gradually increasing testosterone output. These characteristics fit with those outlined above for Leydig cell tumors
- After oophorectomy, the testosterone value fell to 32 ng/dL (1.1 nmol/L), proving that the ovary was indeed the source of testosterone
- The fact that the testosterone was the only abnormality of note misled many about the diagnosis. Some lines of investigation were followed despite existing evidence (either biochemical or in the patient history) militating against them. Searching for congenital adrenal hyperplasia and testicular feminization were such examples

POINTS TO REMEMBER

- Even very high testosterone concentrations are not necessarily associated with major organ dysfunction and virilism in women
- Save time and effort by taking notice of all the information available
- Don't rely on verbal communication with clinicians when a condition has developed over a long time during which different clinicians may have been involved. Formally review the notes
- Confer with colleagues in other specialties (such as histology) and give joint, consistent advice

REFERENCES

1. Young RH, Scully RE. Steroid cell tumours of the ovary. In: Fox H, ed. Obstetrical and gynaecological pathology. London: Churchill Livingstone, 1995:921–31.
2. Langley FA, Fox H. Ovarian tumours; classification, histogenesis and aetiology. In: Fox H, ed. Obstetrical and gynaecological pathology. London: Churchill Livingstone, 1995:727–42.
3. Barth JH. Investigations in the assessment and management of patients with hirsutism. Curr Opin Obstet Gynecol 1997;9;187–92.
4. Marshburn PB, Carr BR. Hirsutism and virilism. A systematic approach to benign and potentially serious causes. Postgrad Med 1995;97:99–106.
5. Lucky AW. Hormonal correlates of acne and hirsutism. Am J Med 1995;98:89S–94S.
6. Griffin JE. Androgen resistance—the clinical and molecular spectrum. N Engl J Med 1992;326:611.

THIS CASE WAS PRESENTED BY DR. BRIAN SENIOR, ROYAL BOLTON HOSPITAL, BOLTON, UK

9. Serum estradiol as a tumor marker: a cautionary tale

PRESENTATION

HISTORY OF PRESENTING COMPLAINT
- A 9-mo-old baby girl was referred to the pediatric endocrine clinic because of breast development, pubic hair growth, and vaginal bleeding

PAST MEDICAL HISTORY
- Normal delivery at term
- No history of note

FAMILY HISTORY
- Second child of nonconsanguineous parents

MEDICATION
- None

ON EXAMINATION
- No skin pigmentation noted
- No thyroid enlargement

WHAT IS THE PROVISIONAL DIAGNOSIS?

WHAT INVESTIGATIONS MIGHT BE REQUESTED?

PROVISIONAL DIAGNOSES
- Gonadotrophin-dependent precocious puberty
- Autonomous estrogen excretion
- Congenital adrenal hyperplasia

INITIAL INVESTIGATIONS AT A TERTIARY REFERRAL LABORATORY
- Initial endocrine biochemistry showed:

Luteinizing hormone	<0.6 mU/mL	(<0.6 U/L)
Follicle-stimulating hormone	<0.8 mU/mL	(<0.8 U/L)
Estradiol	190 pg/mL	(570 pmol/L)
Prolactin	18 ng/mL	(18.0 µg/L)
Cortisol at 0800 h	6.0 µg/dL	(166 nmol/L)

WHAT FURTHER INVESTIGATIONS MIGHT BE CARRIED OUT?

HAVE YOU MADE ANY CHANGES TO YOUR DIAGNOSIS?

FURTHER INVESTIGATIONS

- A gonadotropin hormone–releasing hormone (GnRH) stimulation test showed no change in either luteinizing hormone (LH) or follicle-stimulating hormone (FSH), whereas serum estradiol on two further occasions were 160 pg/mL (563 pmol/L) and 170 pg/mL (590 pmol/L)

Androstenedione	120 ng/dL	(5.4 nmol/L)
Testosterone	40 ng/dL	(1.4 nmol/L)
Dehydroepiandrosterone sulfate	Undetectable	
17α-Hydroxyprogesterone	90 ng/dL	(3.3 nmol/L)

FINAL DIAGNOSIS: Estrogen-secreting ovarian tumor

MANAGEMENT

- An abdominal laparoscopy revealed a right side ovarian tumor 7 cm × 4 cm
- Surgical pathology reported a granulosa cell tumor with extensive luteinization
- Postsurgery estradiol concentrations, measured every 6 mo over a period of several years, remained at <20 pg/mL (<50 pmol/L)—normal prepubertal concentrations
- After a few years the local laboratory decided to measure the 6-monthly estradiol on their newly purchased automated immunoassay analyzer
- The serum estradiol on the analyzer revealed a concentration of 230 pmol/L (70 pg/mL)
- Repeat analysis at the tertiary laboratory continued to measure the estrogen concentration at <50 pmol/L (20 pg/mL) and repeated this finding on subsequent occasions
- The child's father was a local primary-care physician and, therefore, well aware of the implications of a recurrence of excess estrogen secretion

KEY FEATURES

- There is a mild ovarian androgen increase (androstenedione but notably not dehydroepiandrosterone sulfate)
- These results all point to an autonomous source of estrogen secretion
- It is important to distinguish between true isosexual precocious puberty (usually considered to be idiopathic, but with modern imaging techniques, more of them being ascribed to central nervous system lesions), sexual precocity resulting from tumors of the adrenal or ovary (very rarely testicular), and other reasons for signs of premature sexual development. These include McCune-Albright (polyostotic fibrous dysplasia with café-au-lait spots) syndrome, central nervous system tumors, severe hypothyroidism, and congenital adrenal hyperplasia

POINTS TO REMEMBER

- Be aware of the causes of precocious development
- Understand how to use laboratory investigations to diagnose the cause of precocious development
- Understand the interpretation of the results of investigations
- The most significant biochemistry was the very high estradiol concentrations with suppressed LH and FSH, confirmed on GnRH stimulation
- Although GnRH-stimulation tests are frowned upon, they are helpful in confirming the presence of pituitary-driven isosexual precocious puberty
- Judicious selection and interpretation of the appropriate endocrine biochemistry will, in most instances, quickly resolve the underlying cause of the pathology, leading to speedy, relevant, and successful treatment

- Be confident of the capabilities of automated analyzers to measure serum estradiol (and testosterone) at the limits of detection. This is of particular importance when using these assays as tumor markers

REFERENCES
1. Klein KO, Baron J, Barnes KM, Pescovitz OH, Cutler GB. Use of an ultrasensitive recombinant cell bioassay to determine estrogen levels in girls with precocious puberty treated with a luteinizing hormone-releasing hormone agonist. J Clin Endocrinol Metab 1998;83:2387–9.
2. Styne DM, ed. Puberty and its disorders. Endocrinol Metab Clin North Am 1991;20:1–245.
3. Grumbach MM, Styne DM. Puberty: ontogeny, neuroendocrinology, physiology and disorders. In: Wilson JD, Foster DW, Kronenberg HM, Larsen PR, eds. Williams textbook of endocrinology. Philadelphia: WB Saunders, 1998:1509–625.
4. Speiser PW, Susin M, Sasano H, Bohrer S, Markowitz J. Ovarian hyperthecosis in the setting of portal hypertension, J Clin Endocrinol Metab 2000;85:873–7.
5. Young RH, Dickersin GR, Scully RE. Juvenile granulosa cell tumor of the ovary. A clinicopathological analysis of 125 cases. Am J Surg Pathol 1984;8:575–96.
6. White PC, Speiser PW. Congenital adrenal hyperplasia due to 21-hydroxylase deficiency. Endocr Rev 2000;21:245–91.

THIS CASE WAS PRESENTED BY DR. MICHAEL J. DIVER, ROYAL LIVERPOOL UNIVERSITY HOSPITAL, LIVERPOOL, UK

10. A complicated pregnancy

PRESENTATION

HISTORY OF PRESENTING COMPLAINT
- A 35-y-old woman attended her primary-care physician 6 mo after delivery of her third pregnancy
- She complained of a 15-pound-6-ounce (7-kg) weight loss, panic attacks, sweating, racing heart, palpitations, breathlessness, flushing, and vomiting

PAST MEDICAL HISTORY
- She had had two previous large babies weighing 10 pounds 7 ounces (4.74 kg) and 8 pounds 13 ounces (4 kg)
- During her third pregnancy a random serum glucose measured 133 mg/dL (7.4 mmol/L). A glucose tolerance test (GTT) showed a normal fasting glucose concentration of 86 mg/dL (4.8 mmol/L), but after a 75-g glucose load, the concentration at 2 h was 211 mg/dL (11.7 mmol/L)
- A diagnosis of gestational diabetes was made
- She had been treated with a high-fiber, low-fat diet. She did not require insulin
- The pregnancy progressed normally. At 40 wk she delivered a normal healthy large female infant of 10 pounds 11 ounces (4.85 kg) who was normoglycemic
- After delivery she had complained of severe headache and seemed very anxious and agitated
- She developed a soaring blood pressure between 100/80 and 240/120 mmHg. She was treated with a β-blocker and went home
- She returned to the hospital for removal of retained products of conception under general anesthetic on day 10 postpartum. At this time her blood pressure was recorded as 100/80 and 140/100 mmHg
- A plasma glucose was recorded as 256 mg/dL (14.2 mmol/L)
- The whole episode was diagnosed as anxiety-related hypertension and no further action was taken
- Six weeks postpartum a repeat GTT was normal. Her blood pressure was recorded as 120/70 mmHg

ON EXAMINATION
- She looked extremely unwell. She was sweaty, pale, and thin
- Her pulse rate was 120 beats/min and blood pressure was 240/120 mmHg
- She was agitated, anxious, and restless
- She appeared dehydrated
- She was admitted to hospital

WHAT IS YOUR PROVISIONAL DIAGNOSIS OF HER PRESENT COMPLAINT?

WHAT INVESTIGATIONS WOULD YOU REQUEST?

WHAT IS THE DEFINITION OF GESTATIONAL DIABETES?

WHY SHOULD DIABETES BE LOOKED FOR AND TREATED IN PREGNANCY?

PROVISIONAL DIAGNOSES
- Diabetes mellitus
- Severe thyrotoxicosis
- Anemia
- Severe infection
- Pheochromocytoma

INITIAL INVESTIGATIONS

There was 3+ glucose in her urine

Random blood glucose	141 mg/dL	(7.8 mmol/L)
Serum		
Thyrotropin	1.68 µU/mL	(1.68 mIU/L)
Free thyroxine	1.27 ng/dL	(16.3 pmol/L)
Hemoglobin	12.0 g/dL	(120 g/L)
Leukocyte count	$7.4 \times 10^3/\mu L$	($7.4 \times 10^9/L$)
Urine		
4-Hydroxy-3-methoxymandelic acid (HMMA)	5.7 µg/24 h	(33.8 µmol/24 h)
Homovanillic acid	4.46 mg/24 h	(24.5 µmol/24 h)
Adrenaline	18.3 µg/24 h	(100 µmol/24 h)
Noradrenaline	881 µg/24 h	(5209 µmol/24 h)
Dopamine	342 µg/24 h	(2236 µmol/24 h)

GESTATIONAL DIABETES
- Gestational diabetes is diabetes mellitus induced by pregnancy but resolving at the end of gestation
- The diagnostic criteria for gestational diabetes was adopted by the 4th International Workshop-Conference on Gestational Diabetes Mellitus in 1998. It is based on an oral 75-g GTT administered between weeks 24 and 28 of gestation
- Patients with a fasting glucose >94 mg/dL (>5.2 mmol/L), 1-h glucose >178 mg/dL (>9.9 mmol/L), and 2-h glucose >200 mg/dL (>11.1 mmol/L) or a fasting glucose >124 mg/dL (>6.9 mmol/L) are deemed to have type 1 or type 2 diabetes mellitus. In other centers, a 100-g glucose load 3-h test is performed

WHY SHOULD DIABETES MELLITUS BE LOOKED FOR AND TREATED IN PREGNANCY?
- There are problems for both mother and fetus associated with a raised blood glucose: Polyhydramnios, pre-eclampsia, and maternal infection are increased for the mother
- There is an increased likelihood of a large baby weighing >8 pounds 13 ounces (4 kg), increased risk of intrauterine death, and increased incidence of obstructed labor and neonatal hypoglycemia

---

WHAT FURTHER INVESTIGATIONS, IF ANY, WOULD YOU REQUEST?

HAVE YOU MADE ANY CHANGES TO YOUR PROVISIONAL DIAGNOSIS?

A COMPLICATED PREGNANCY

You wish to confirm adrenal tumor and exclude multiple endocrine neoplasia 2 (MEN-2)
- Computed tomography (CT) scan of adrenals
- Serum calcium and parathyroid hormone (PTH)
- Calcitonin

RESULTS

- CT scan showed a 4-cm tumor in the left adrenal
- Serum calcium   8.96 mg/dL   (2.24 mmol/L)
- PTH   23.9 pg/mL   (23.9 ng/L)
- Calcitonin   <14 ng/L   (<0.4 pmol/L)

WHAT ARE THE INDICATIONS TO LOOK FOR IN PHEOCHROMOCYTOMA?

- Hypertensive crisis occurring during induction of anesthesia, surgery, or labor
- Hypertension existing with headache and hyperhidrosis
- Family history of pheochromocytoma
- Other endocrine or neural disorders
- Normotensive or hypertensive patients with unexplained abdominal or chest pain

FINAL DIAGNOSIS: Pheochromocytoma

MANAGEMENT

- The tumor was surgically removed under close anesthetic supervision. The tumour weighed 28 g, and histological examination confirmed it to be a pheochromocytoma
- She made a full recovery and has remained normotensive

KEY FEATURES

PRESENTATION

- Gestational diabetes occurs with an incidence of 0.5–1% of pregnancies. About 0.8% of pregnant women show glucose in the second fasting morning urine sample, and 15% of these have chemical diabetes. The renal threshold for glucose decreases in pregnancy, but it does not indicate diabetes
- This patient's symptoms were classic of pheochromocytoma: hypertension, headache, and hyperhidrosis (the 3Hs). Although it is regarded as a rare tumor, data from the Mayo Clinic suggest it is the cause of hypertension in 1:1000 to 1:2500 patients. It is considerably underdiagnosed, with 40% of cases only diagnosed at autopsy. From the onset of symptoms to diagnosis is usually 4.5 y. Death can be due to a hypertensive crisis resulting in myocardial infarction, left ventricular failure, pulmonary edema, cerebrovascular accident, or ruptured aortic aneurysm. Death may also occur during anesthesia due to cardiac dysfunction
- Pheochromocytoma in pregnancy is associated with 48% mortality in the mother and 55% mortality in the fetus. Symptoms increase during the third trimester due to outpouring of catecholamines from the tumor under pressure from the enlarging fetus. The most dangerous time is from the onset of labor to 48–72 h postpartum. Pheochromocytoma in pregnancy may be misdiagnosed as pre-eclampsia

PATHOPHYSIOLOGY
- During the 9 mo of pregnancy, there are large changes in hormone production, and pregnancy is described as an insulin-resistant state. This is due in part to insulin utilization in the placenta, which contains insulin receptors and has insulinase activity
- Insulin concentrations in the serum of pregnant women with normal blood glucose are raised in the second trimester compared with their nonpregnant sisters
- Human placental lactogen, a peptide produced by the placenta, induces lipolysis and reduces sensitivity to insulin
- Glucose intolerance is a feature of pheochromocytoma and is caused by the action of noradrenaline and adrenaline on the α_1 and α_2 adrenoreceptors. They stimulate glycogenolysis and inhibit insulin secretion. This action is similar to the action of human placental lactogen and insulinase in the placenta

DIAGNOSIS
- HMMA is frequently used as a screening test in the urine to detect excess production of catecholamines (see figure). It may not be the best test to use
- There is considerable evidence that noradrenaline and adrenaline measured by high-performance liquid chromatography are better markers of excess production
- Because hormone production is episodic, a urine collection is preferred to a plasma sample, and several urines may be required before the diagnosis is discarded

FIGURE METABOLIC PATHWAY FOR THE PRODUCTION OF CATECHOLAMINES
VMA, 4-hydroxy-3-methoxymandelic acid; HMMA, 3-methoxy-4-hydroxymandelic acid; DOPA, l-3,4-dihydroxyphenylalanine

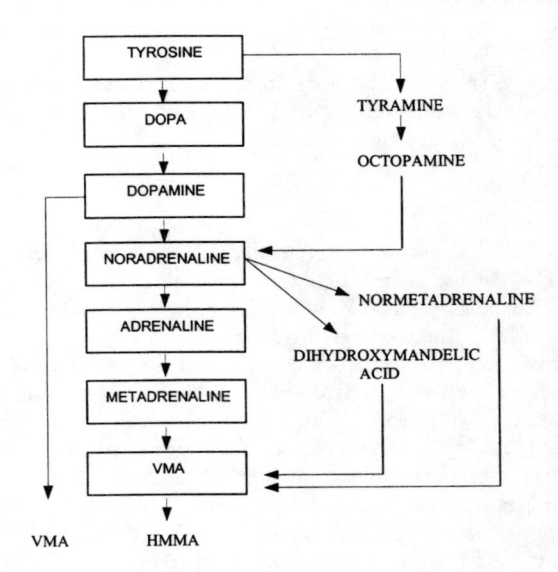

```
TYROSINE ──────────────┐
   │                   ▼
   ▼                TYRAMINE
  DOPA                 │
   │                   ▼
   ▼               OCTOPAMINE
DOPAMINE
   │
   ▼
NORADRENALINE ──────► NORMETADRENALINE
   │              ──► DIHYDROXYMANDELIC ACID
   ▼
ADRENALINE
   │
   ▼
METADRENALINE
   │
   ▼
  VMA
   │
   ▼
  HMMA

VMA
```

POINTS TO REMEMBER
- Glycosuria occurs in pregnancy due to lowered renal threshold for glucose
- Pregnancy is an insulin-resistant state
- Hypertension with glucose intolerance should always prompt an investigation for pheochromocytoma. Catecholamines cause glucose intolerance by increasing glycogenolysis and inhibiting insulin release
- Tumor hormone production is episodic. Catecholamines and their metabolites should all be measured in an acidified urine before the diagnosis is discarded
- Pheochromocytoma in pregnancy can be fatal to both mother and baby

REFERENCES
1. Garner PR. Glucose metabolism in pregnancy. Clin Biochem 1995;28:499–502.
2. Kjos Siri L, Buchanon TYA. Gestational diabetes mellitus. N Engl J Med 1999;341:1749–56.
3. Peaston RT, Lai LC. Biochemical detection of phaemochromocytoma: should we still be measuring urinary HMMA? J Clin Pathol 1993;46:734–7.
4. Ross EJ, Griffith DNW. The clinical presentation of phaeochromocytoma. Q J Med 1989;71:485–96.
5. Stewart MF, Reed P, Weincove C, Moriarty KJ, Ralston AJ. Biochemical diagnosis of phaeochromocytoma: two instructive case reports. J Clin Pathol 1993;46:280–5.

THIS CASE WAS PRESENTED BY DR. HAZEL WILKINSON AND DR. PAUL E. JENNINGS, YORK DISTRICT HOSPITAL, YORK, UK

11. A complication of anticonvulsant therapy

PRESENTATION

HISTORY OF PRESENTING COMPLAINT
- A 45-y-old man with epilepsy was at his annual outpatient clinic visit
- He complained of muscle pain in his legs that became more severe when walking for more than about a quarter of a mile
- He had experienced the pain for the past few weeks as far as he could remember

DIRECT QUESTIONING
- He reported no evidence of symptoms of epileptic fits over the past 8 y
- The patient led a normal lifestyle, which he described as fairly sedentary
- He did not take exercise specifically and was not aware of any precipitating factor before becoming aware of the pains in his legs

PAST MEDICAL HISTORY
- The patient had been diagnosed as epileptic 12 y previously at the age of 33 y
- He had subsequently been put on phenytoin therapy and attended an outpatient clinic annually
- Four years ago he had been admitted to hospital after feeling unwell for several days. He remembered his upper abdomen being tender to the touch and that the doctor had told him that he had biliary colic
- Subsequent annual visits to the outpatient clinic had been routine

SOCIAL HISTORY
- Nothing of note

FAMILY HISTORY
- Nothing of note

ON EXAMINATION
- He appeared to be of normal stature and build; weight 172 pounds (78 kg), height 5 feet 10 inches (178 cm), body mass index 24.4 kg/m^2
- He appeared to be alert
- There was no evidence of abdominal tenderness, and his sclera were of normal color

---

WHAT IS YOUR PROVISIONAL DIAGNOSIS?

WHAT INVESTIGATIONS WOULD YOU REQUEST?

PROVISIONAL DIAGNOSIS
- Known epilepsy
- Past history of:
    Gallstones
- Myositis
- Complications of long-term treatment with phenytoin

INITIAL INVESTIGATIONS
Serum

Sodium	141 mEq/L	(141 mmol/L)
Potassium	3.9 mEq/L	(3.9 mmol/L)
Bicarbonate	27 mEq/L	(27 mmol/L)
Urea	10 mg/dL	(3.6 mmol/L)
Creatinine	0.8 mg/dL	(71 µmol/L)
Calcium	8.32 mg/dL	(2.08 mmol/L)
Phosphate	6.2 mg/dL	(2.0 mmol/L)
Alkaline phosphatase	285 IU/L	(285 IU/L)
Alanine aminotransferase	27 IU/L	(27 IU/L)
Bilirubin	0.6 mg/dL	(10 µmol/L)
Albumin	4.0 g/dL	(40 g/L)
Total protein	7.4 g/dL	(74 g/L)
Creatine kinase	115 IU/L	(115 IU/L)
Phenytoin	2.2 mg/dL	(85 µmol/L)

WHAT FURTHER INVESTIGATIONS, IF ANY, WOULD YOU REQUEST?

HAVE YOU MADE ANY CHANGES TO YOUR PROVISIONAL DIAGNOSIS?

WHAT IS YOUR DIAGNOSIS?

A COMPLICATION OF ANTICONVULSANT THERAPY

WORKING DIAGNOSES
- Epileptic on anticonvulsant therapy
- Hypocalcemia
- Metabolic bone disease as a secondary consequence of long-term anticonvulsant therapy

FURTHER INVESTIGATIONS
- A review of all the laboratory data in the patient's notes showed two previous instances of an elevated serum alkaline phosphatase. There had been one previous request for a serum γ-glutamyltransferase, which was elevated. At the time of the episode of abdominal discomfort there had been a slight elevation of the bilirubin and alanine aminotransferase concentrations. The data are summarized in the table below
- Confirmation of the effect of phenytoin on enzyme induction was sought, with a γ-glutamyltransferase activity of 360 IU/L found
- The source of the elevated alkaline phosphatase was determined initially by electrophoresis, which showed that the major isoform found was that derived from bone. Subsequent radioimmunoassay for the bone isoform found a concentration of 76 μg/L (reference range in adults 7–28 μg/L)
- Bone mineral density measurements found values at the lower limit of the age-related reference range for the femoral neck and trochanter region and within the reference range for the spine
- Assessment of vitamin D status with a measurement of serum 1,25-dihydroxycholecalciferol found a value of 3.2 ng/mL (9 nmol/L)
- The urinary 24-h calcium excretion was found to be 44 mg/24 h (1.1 mmol/ 24 h)
- The serum parathyroid hormone concentration was found to be 140 pg/mL (140 ng/L)

TABLE
Serum biochemical parameters measured at various times since diagnosis of epilepsy

	FIRST VISIT: EPILEPSY DIAGNOSED	1 Y LATER: ROUTINE	4 Y LATER: ROUTINE	8 Y LATER: BILIARY COLIC	12 Y LATER: ROUTINE, MUSCLE PAIN
Bilirubin [mg/dL (μmol/L)]	07 (12)	0.8 (14)	0.6 (10)	1.8 (30)	0.6 (10)
ALT (IU/L at 37 °C)	18	23	19	50	27
ALP (IU/L at 37 °C)	96	109	109	142	285
GGT (IU/L at 37 °C)	-	-	-	-	250
Calcium [mg/dL (mmol/L)]	-	-	-	10.1 (2.5)	8.35 (2.08)
Albumin [g/dL (g/L)]	4.2 (42)	4.2 (42)	4.3 (43)	4.1 (41)	4.0 (40)

ALT, alanine aminotransferase; ALP, alkaline phosphatase; GGT, γ-glutamyltransferase

FINAL DIAGNOSIS: Secondary metabolic bone disease as a consequence of long-term treatment with phenytoin

MANAGEMENT
- He was treated with vitamin D and calcium supplements
- There was an improvement in the serum calcium and 1,25-dihydroxycholecalciferol concentrations

KEY FEATURES

PRESENTATION
- Hypocalcemia, hypocalciuria, or a reduced concentration of serum 1,25-dihydroxycholecalciferol may be the first sign of altered calcium homeostasis and may only be found if patients on phenytoin are screened for these tests as part of their routine review
- Muscle weakness and particularly muscle pain and difficulty with walking are a sign of severe 1,25-dihydroxycholeciferol deficiency
- Bone loss is known to accompany treatment with the antiepileptic drugs phenytoin and carbamazepine
- Bone turnover is accelerated independently of vitamin D status

BIOCHEMISTRY
- The bone disease is generally considered to be due to treatment-induced vitamin D deficiency
- This may be the result of an accelerated conversion of vitamin D and 25-hydroxy-cholecalciferol to inactive polar metabolites by drug-induced hepatic microsomal enzymes
- It has been suggested that there might be a direct effect of antiepileptic drugs on bone cells
- A reduced level of calcium absorption has been demonstrated
- The hypocalcemia is thought to be responsible for a secondary hyperparathyroidism

DIAGNOSIS
- Routine monitoring of the serum calcium and alkaline phosphatase can provide an effective monitoring strategy backed up by urinary calcium excretion
- 25-Hydroxy- or 1,25-dihydroxycholecalciferol and parathyroid hormone concentrations will confirm the vitamin D and parathyroid gland status
- Bone mineral density measurements may be particularly useful in women, who may be at greater risk of fracture, particularly in later life
- In this situation the serum γ-glutamyltransferase activity cannot be used to deduce the source of an alkaline phosphatase elevation due to the effect of the antiepileptic drug on enzyme induction

TREATMENT
- Vitamin D supplementation
- Possibly calcium supplementation initially

POINTS TO REMEMBER

- Some antiepileptic drugs induce the de novo synthesis of hepatic microsomal enzymes
- Antiepileptic drugs may produce jaundice in some patients. Phenytoin tends to produce a predominantly hepatocellular jaundice
- The enzyme induction may result in a relative vitamin D deficiency leading to hypocalcemia
- The source of a raised alkaline phosphatase may be hepatic or osteoblastic. The γ-glutamyltransferase activity cannot be used to differentiate the source in this situation because the enzyme will be induced by the antiepileptic drug, that is, not indicating a cholestatic drug reaction
- Thus, you need to use a method that differentiates between the major isoforms of alkaline phosphatase, such as electrophoresis, wheat germ lectin precipitation, or immunoassay for the bone isoform

REFERENCES

1. Ala-Houhala M, Korpela R, Koivikko M, Koskinen T, Koskinen M, Koivula T. Long-term anticonvulsant therapy and vitamin D metabolism in ambulatory pubertal children. Neuropediatrics 1986;17:212–6.
2. Gough H, Goggins T, Bissessar A, Baker M, Crowley M, Callaghan N. A comparative study of the relative influence of different anticonvulsant drugs, UV exposure and diet on vitamin D and calcium metabolism in out-patients with epilepsy. Q J Med 1986;59:569–77.
3. Hunt PA, Wu-Chen ML, Handal NJ, et al. Bone disease induced by anticonvulsant therapy and treatment with calcitriol (1,25-dihydroxyvitamin D3). Am J Dis Child 1986;140:715–8.
4. Maeda K, Ikeda H. High 1,25-dihydroxyvitamin D concentrations in plasma in patients receiving antiepileptic drugs. Jpn J Psychiatry Neurol 1986;40:57–60.
5. Nishiyama S, Kuwahara T, Matsuda I. Decreased bone density in severely handicapped children and adults, with reference to the influence of limited mobility and anticonvulsant medication. Eur J Pediatr 1986;144:457–63.
6. Timperlake RW, Cooke SD, Thomas KA, et al. Effects of anticonvulsant drug therapy on bone mineral density in a pediatric population. J Pediatr Orthop 1988;8:467–70.
7. Valimaki MJ, Tiihonen M, Laitinen K, et al. Bone mineral density measured by dual-energy X-ray absorptiometry and novel markers of bone formation and resorption in patients on antiepileptic drugs. J Bone Miner Res 1994;9:631–7.

THIS CASE WAS PRESENTED BY PROFESSOR CHRISTOPHER P. PRICE, ROYAL LONDON HOSPITAL, BARTS AND LONDON NHS TRUST, LONDON, UK

12. A complication of a road traffic accident

PRESENTATION

HISTORY OF PRESENTING COMPLAINT
- A 53-y-old man presented after a road traffic accident. No other party was involved, and he had suffered a loss of consciousness
- This accident was preceded by the development of increasingly severe headaches over a period of 3 d before the accident
- His general health was otherwise good

PAST MEDICAL HISTORY
- The patient had had a cerebral artery surgically clipped at the age of 39 y
- He was a smoker, but there was no history of alcohol abuse, diabetes, hypertension, or peripheral vascular or respiratory disease
- He took no medication

ON EXAMINATION
- He was of normal stature and build
- He was alert with a Glasgow Coma Score of 15/15. There was mild neck stiffness but no localizing neurological signs
- He was well hydrated with blood pressure of 135/80 mmHg

INITIAL MANAGEMENT
- The history of loss of consciousness and trauma prompted a computed tomography scan, which confirmed a subarachnoid hemorrhage associated with a large aneurysm of the left anterior communicating artery. This was confirmed by subsequent cerebral angiography. He was treated surgically, but 5 d post-op, he was noted to have become hyponatremic with a fall in plasma sodium from 135 to123 mEq/L (135 to 123 mmol/L)

WHAT IS YOUR PROVISIONAL DIAGNOSIS?

WHAT INVESTIGATIONS WOULD YOU REQUEST?

PROVISIONAL DIAGNOSIS

- All causes of hyponatremia should be considered at this stage

INITIAL INVESTIGATIONS

	FLUID INPUT (mL)	FLUID OUTPUT (mL)	PLASMA SODIUM (mEq/L, mmol/L)	PLASMA POTASSIUM (mEq/L, mmol/L)	PLASMA CREATININE [mg/dL (µmol/L)]	URINE SODIUM (mEq/L, mmol/L)	URINE OSMOLALITY (mOsm/kg)
Pre-op	–	–	135	4.0	0.8 (75)	–	–
Day 1	2950	2800	132	3.6	0.9 (84)	–	–
Day 2	–	–	133	3.1	0.8 (72)	–	–
Day 3	–	–	132	3.5	0.9 (76)	–	–
Day 4	5050	4250	127	3.3	0.8 (67)	146	486
Day 5	2400	3100	123	3.4	0.8 (74)	142	507

Total thyroxine 5.0 µg/dL (64 nmol/L)
Total triiodothyronine 71 ng/dL (1.1 nmol/L)
Thyroid-stimulating hormone 2.23 µU/mL (2.23 mU/L)

A tentative diagnosis of syndrome of inappropriate secretion of antidiuretic hormone (SIADH) was made, and he was started on a fluid-restriction regime

---

WHAT FURTHER INVESTIGATIONS, IF ANY, WOULD YOU REQUEST?

HAVE YOU MADE ANY CHANGES TO YOUR PROVISIONAL DIAGNOSIS?

WHAT IS YOUR DIAGNOSIS?

- He was noted to be underhydrated (blood pressure supine 125/74 mmHg and sitting 116/73 mmHg) with severe polyuria and natriuresis
- He was treated with normal saline (3 L/d) and started on hydrocortisone pending assessment of adrenal reserve
- Despite this management he remained in negative fluid balance

	FLUID INPUT (mL)	FLUID OUTPUT (mL)	PLASMA SODIUM (mEq/L, mmol/L)
Day 6	3175	6200	125
Day 7	6325	7600	130
Day 8	3970	4345	125
Day 9	–	–	124
Day 10	2600	5300	124

FINAL DIAGNOSIS: Cerebral salt wasting (CSW)

MANAGEMENT
- An adequate adrenal reserve was confirmed using an intramuscular glucagon test. Glucagon was used because it tests the entire pituitary-adrenal axis, which is essential in this case because CSW is related to hypothalamic damage
- After the diagnosis, he was started on oral sodium [100 mEq/24 h (100 mmol/24 h)] in addition to saline infusion, and over the next 7 d, the inappropriate diuresis ceased and his plasma sodium rose from 125 to 134 mEq/L (125 to 134 mmol/L)

KEY FEATURES
- Hyponatremia, polyuria, severe natriuresis [day 4, sodium loss was 621 mEq/24 h (621 mmol/24 h)]

PRESENTATION
- CSW can complicate all cases of hypothalamic damage, such as craniopharyngioma, aneurysms of the communicating arteries of the circle of Willis

BIOCHEMISTRY
- Patients with subarachnoid hemorrhage have raised concentrations of the brain natriuretic peptide, which appears to be raised in proportion to the intracranial pressure. This is associated with a suppression of the normal salt and water homeostatic mechanism: both plasma ADH and the renin-aldosterone axis are suppressed

DIAGNOSIS
- The diagnosis should be considered in all cases of cerebral injury because plasma volume loss >10% occurs in 50% of patients with aneurysmal subarachnoid hemorrhage
- The important clinical features that differentiate CSW from SIADH are reduced plasma volume as shown by reduced central venous pressure and weight loss. The laboratory investigations will reveal features of dehydration and raised hematocrit and plasma urea-creatinine ratio together with inappropriately elevated urine sodium concentration
- The response of hyponatremia and volume contraction to saline infusion is prolonged with CSW, whereas the correction of hyponatremia is short lived in cases of SIADH

TREATMENT
- The objectives of treatment are volume replacement and the maintenance of positive sodium balance. Care should be taken not to allow the plasma sodium to rise too quickly in order to reduce the risk of precipitating pontine demyelination

POINTS TO REMEMBER
- Hyponatremia after subarachnoid hemorrhage may be due to secretion of brain natriuretic peptide and the compensatory salt and water recovery mechanism's being paralyzed by inhibition of both antidiuretic hormone and renin-aldosterone secretion
- It is important to differentiate hyponatremia due to CSW from SIADH. CSW causes fluid depletion, whereas SIADH does not. Fluid restriction is the first line of treatment for SIADH, but this can exacerbate vasospasm and produce ischemia and infarction in patients with salt wasting because they are likely to be already dehydrated

REFERENCES
1. Berendes E, Walter M, Cullen P, et al. Secretion of brain natriuretic peptide in patients with aneurysmal subarachnoid hemorrhage. Lancet 1997;349:245–9.
2. Harrigan MR. Cerebral salt wasting—a review. Neurosurgery 1996;38:152–60.
3. Isotani E, Suzuki R, Tomita K, et al. Alterations in plasma concentrations of natriuretic peptides and antidiuretic hormone after subarachnoid hemorrhage. Stroke 1994;25:2198–203.
4. Zafonte RD, Mann NR. Cerebral salt wasting syndrome in brain injury patients: a potential cause of hyponatremia. Arch Phys Med Rehabil 1997;78:540–2.

THIS CASE WAS PRESENTED BY DR. JULIAN H. BARTH AND DR. PAUL E. BELCHETZ, LEEDS GENERAL INFIRMARY, LEEDS, UK

13. Chronic hyponatremia

PRESENTATION

HISTORY OF PRESENTING COMPLAINT
- A 66-y-old woman was admitted for investigation of long-standing hyponatremia
- The hyponatremia was documented over the previous 2 y with serum sodium ranging between 119 and 131 mEq/L (119 and 131 mmol/L)
- There was no history of weight loss, lung disease, or central nervous system disease
- She complained of tiredness in the evening

PAST MEDICAL HISTORY
- She had suffered type 1 diabetes for 44 y. Her diabetic control was moderate, with home glucose readings between 90 and 216 mg/dL (5 and 12 mmol/L). HbA$_{1C}$ was 8.5% (Diabetes Control and Complications Trial standardized)
- There was persistent evidence of microalbuminuria with albumin-creatinine ratios on early morning urines of between 64 and 100 mg/g (7.2 and 11.3 mg/mmol). The serum creatinine ranged between 1.2 and 1.4 mg/dL (104 and 122 µmol/L)
- Serum potassium measurements ranged between 4.6 and 5.2 mEq/L (4.6 and 5.2 mmol/L)
- Other treated medical conditions were vitamin B$_{12}$ deficiency and asthma
- Medication consisted of insulin (Human Mixtard) and vitamin B$_{12}$ injections

SOCIAL HISTORY
- Lives at home with her husband
- Nonsmoker, no alcohol consumed

ON EXAMINATION
- Weight 106 pounds (48 kg), height 5 feet 0 inches (1.52 m), body mass index 20.8
- She was fully in charge of all of her faculties with no evidence of mental impairment or confusion
- Clinically normovolemic
- Blood pressure 130/80 mmHg
- No evidence of diabetic neuropathy
- Fundi normal

WHAT IS YOUR PROVISIONAL DIAGNOSIS?

WHAT INVESTIGATIONS WOULD YOU REQUEST?

PROVISIONAL DIAGNOSES
- Pseudohyponatremia due to hypertriglyceridemia associated with diabetes mellitus
- Hypothyroidism
- Secondary hypoadrenalism
- Hyporeninemic hypoaldosteronism
- Chronic hyponatremia due to impaired water excretion [syndrome of inappropriate antidiuretic hormone secretion (SIADH)]

INITIAL INVESTIGATIONS

Serum
Sodium	124 mEq/L	(124 mmol/L)
Potassium	4.6 mEq/L	(4.6 mmol/L)
Urea	16 mg/dL	(5.6 mmol/L)
Creatinine	1.2 mg/dL	(109 mmol/L)
Bicarbonate	25 mEq/L	(25 mmol/L)
Osmolality	264 mOsm/kg	(264 mOsm/kg)
Glucose	151 mg/dL	(8.4 mmol/L)
Free thyroxine	1.06 ng/dL	(13.7 pmol/L)
Thyroid-stimulating hormone	1.2 µU/mL	(1.2 mU/L)
Aldosterone	8.7 ng/dL	(240 pmol/L)
Renin (plasma renin activity)	<0.24 ng/h/mL	(<0.2 pmol/h/mL)

Short Synacthen (cosyntropin) test:
0-min cortisol	14 µg/dL	(380 nmol/L)
30-min cortisol	28 µg/dL	(785 nmol/L)

Urine
Sodium	95 mEq/L	(95 mmol/L)
Potassium	28 mEq/L	(28 mmol/L)
Urea	2.5 g/L	(90 mmol/L)
Osmolality	359 mOsm/kg	(359 mOsm/kg)

Chest X-ray — Normal

WHAT FURTHER INVESTIGATIONS, IF ANY, WOULD YOU REQUEST?

HAVE YOU MADE ANY CHANGES TO YOUR PROVISIONAL DIAGNOSES?

CHRONIC HYPONATREMIA

WORKING DIAGNOSIS
Chronic hyponatremia due to impaired water excretion (SIADH). In this patient the diagnosis does not have a clear etiology, but a management strategy is required. The key questions are:
- Does any degree of osmoregulatory control exist?
- What is the current daily urine volume and osmotic load?

FURTHER INVESTIGATIONS
Water load–deprivation test
- Water load 20 mL/kg consumed at time 0

TIME AFTER WATER LOAD (h)	SERUM OSMOLALITY (mOsm/kg)	URINE VOLUME (mL)	URINE OSMOLALITY (mOsm/kg)
1	271	150	362
2	–	97	358
3	263	95	374
4	–	62	376
5	–	142	386
6	–	56	383
7	265	125	384

- 24-h urine collection
  Volume 1664 mL

Sodium	99 mEq/L	(99 mmol/L)
	65 mEq/24 h	(65 mmol/24 h)
Potassium	25 mEq/L	(25 mmol/L)
	43 mEq/24 h	(42 mmol/24 h)
Urea	2.3 g/L	(82.7 mmol/L)
	3.9 g/24 h	(138 mmol/24 h)
Osmolality	360 mOsm/kg	(360 mOsm/kg)
	599 mOsm/24 h	(599 mOsm/24 h)

MANAGEMENT
- This woman has essentially fixed urine concentration around osmolality 350–400 mOsm/kg (350–400 mOsm/kg)
- Because she cannot dilute her urine, the urine flow depends on osmotic intake (see figure)
- If the osmotic load for excretion is always ~ 600 mOsm/24 h (~600 mOsm/24 h), then due to the essentially fixed urine concentration, the obligatory urine volume is ~1.6 L
- If fluid intake exceeds 1.6 L, then this water will, in the short term, be retained and hyponatremia will become more severe
- Management is to prevent further reduction in serum sodium concentration because of the risks of impaired cerebral function
- Control of serum sodium concentration may be achieved either through strict control of fluid intake, or by increasing the osmotic load for excretion, or by combining both stratagems

FIGURE   RELATIONSHIP OF URINE VOLUME TO THE PREVAILING URINE OSMOLALITY FOR DAILY OSMOTIC LOADS FOR EXCRETION OF 600 (●) AND 1200 (■) mOsm

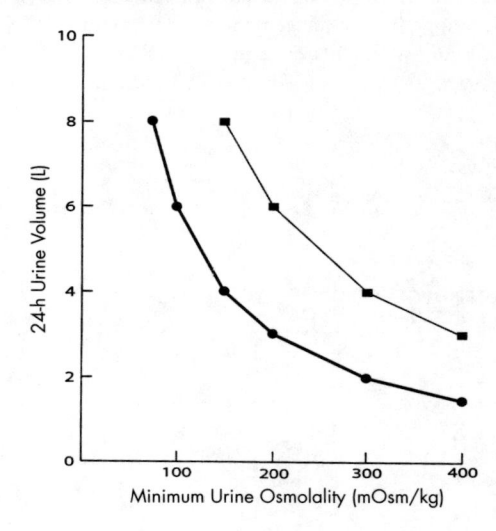

POINTS TO REMEMBER

- The essential differentiation of hyponatremia into hypovolemic, hypervolemic, and normovolemic forms is a clinical decision based on clinical examination and basic urea and electrolyte measurements in serum and urine
- Care should be exercised in differentiating the acute onset of normovolemic hyponatremia (usually iatrogenic with documented duration of <48 h) from a chronic onset; the management of acute and chronic forms is completely different
- Chronic normovolemic hyponatremia is common in the elderly, and for the majority who are clinically stable and without symptoms the etiology remains idiopathic; extensive routine diagnostic procedures are normally not warranted
- It is imperative that hypothyroidism and secondary hypoadrenalism are excluded specifically and early
- Management of chronic normovolemic hyponatremia (when no primary treatable condition is evident) should aim to prevent further reduction in serum sodium because morbidity increases as serum sodium concentration falls: care must be exercised in balancing fluid and osmotic input to the limitation of the capacity to excrete water

REFERENCES

1. Adrogue HJ, Madias NE. Hyponatremia. N Engl J Med 2000;342:1581–9.
2. Hirshberg B, Ben-Yehuda A. The syndrome of inappropriate antidiuretic hormone secretion in the elderly. Am J Med 1997;103:270–3.
3. Black RM. Diagnosis and management of hyponatremia. J Intensive Care Med 1989;4:205–20.
4. Bartter FC, Schwartz WB. The syndrome of inappropriate secretion of antidiuretic hormone. Am J Med 1967;42:790–806.

THIS CASE WAS PRESENTED BY DR. MICHAEL D. PENNEY, ROYAL GWENT HOSPITAL, NEWPORT, WALES

14. A curious case of hyperkalemia

PRESENTATION

HISTORY OF PRESENTING COMPLAINT
- A family doctor called the laboratory for advice about a 75-y-old woman who had had a persistently raised serum potassium concentration over the previous 6 mo

PAST MEDICAL HISTORY
- The patient had had an anterior septal myocardial infarction 18 mo previously, for which she received thrombolytic treatment and was started on captopril
- The examination at that time showed a pulse rate of 120/min and blood pressure of 120/70 mmHg
- The echocardiogram showed a dilated left ventricle and akinetic septum
- She was mobilized and discharged on aspirin 75 mg every day and captopril. The recommendation in the discharge report was that as long as her systolic blood pressure remained >110 mmHg, the captopril should be gradually increased from 6.25 mg twice a day with regular monitoring of urea and electrolytes
- The chest X-ray and routine biochemical profile (including potassium) were completely normal at discharge

SOCIAL HISTORY
- Lives alone
- History of tobacco use—20 cigarettes per day for 35 y
- History of alcohol use—<1 unit per week

CURRENT MEDICATIONS
- Frusemide 20 mg every day
- Isosorbide 40 mg twice a day
- Aspirin 75 mg every day
- Captopril 25 mg twice a day

ON EXAMINATION
- The patient was a slight woman with a ruddy complexion.
- The examination of the cardiovascular system suggested left ventricular failure. Blood pressure was 120/90 mmHg; pulse rate was 84 beats/min and regular
- The liver edge was palpable at the right costal margin
- The spleen tip was felt 4 cm below the left costal margin
- No other abnormalities were found

WHAT IS YOUR PROVISIONAL DIAGNOSIS?

WHAT INVESTIGATIONS WOULD YOU REQUEST?

INITIAL INVESTIGATIONS

- It was thought that captopril may be the cause of this patient's hyperkalemia.
- Captopril was stopped for 1 wk and a fresh sample was collected

Serum		
Sodium	140 mEq/L	(140 mmol/L)
Potassium	5.6 mEq/L	(5.6 mmol/L)
Bicarbonate	30 mEq/L	(30 mmol/L)
Urea	17 mg/dL	(6.1 mmol/L)
Creatinine	1.4 mg/dL	(121 µmol/L)
Total protein	6.4 g/dL	(64 g/L)
Albumin	4.2 g/dL	(42 g/L)
Calcium	9.8 mg/dL	(2.44 mmol/L)
Hemoglobin	17.0 g/dL	(170 g/L)
Leukocyte count	$16.7 \times 10^3/\mu L$	(16.7×10^9/L)
Platelets	$1200 \times 10^3/\mu L$	(1200×10^9/L)
Erythrocyte count	$6.19 \times 10^6/\mu L$	(6.19×10^{12}/L)
Packed cell volume	52.3%	(0.523)
Mean cell volume	84.5 µm^3	(84.5 fL)
Mean cell hemoglobin	27.5 pg	
Mean cell hemoglobin concentration	32.5 gHb/dL	325 gHb/L
Erythrocyte distribution width	17.8%	
Mean platelet volume	10.3 µm^3	(10.3 fL)
Platelet distribution width	17.8	
Neutrophils	$13.2 \times 10^3/\mu L$	(13.2×10^9/L)
Erythrocyte sedimentation rate	7 mm/h	

WHAT FURTHER INVESTIGATIONS, IF ANY, WOULD YOU REQUEST?

HAVE YOU MADE ANY CHANGES TO YOUR PROVISIONAL DIAGNOSIS?

WHAT IS YOUR DIAGNOSIS?

A CURIOUS CASE OF HYPERKALEMIA

PROVISIONAL DIAGNOSIS

- Hyperkalemia can be due to excessive intake or decreased loss of potassium. A normal intake may be excessive if excretion is decreased, for example, in renal failure. Excessive intake is otherwise almost always parenteral and iatrogenic. Decreased loss of potassium can occur in renal disease and in conditions associated with low aldosterone concentrations, most notably, Addison's disease and treatment with an angiotensin-converting enzyme (ACE) inhibitor. Hyperkalemia can also result from a redistribution of potassium from the intra- to the extracellular compartment, which occurs in acidosis, for example, diabetic ketoacidosis
- Captopril is an ACE inhibitor and therefore inhibits the conversion of angiotensin I to angiotensin II. This inhibition in turn causes a reduction in circulating aldosterone. The main action of aldosterone is to stimulate sodium resorption in the distal convoluted tubules in the kidney in exchange for potassium and hydrogen ions. Thus, a reduction of aldosterone results in increased loss of sodium and retention of potassium
- Spurious hyperkalemia due to leakage of potassium from blood cells often occurs in vitro. If hyperkalemia is found unexpectedly, the possibility that it is spurious should be eliminated by repeating the measurement on a fresh sample

FURTHER INVESTIGATIONS

- A fresh plasma sample was requested and the results are shown below:
  Plasma
  Specimen comment: FROM HEPARIN

Sodium	142 mEq/L	(142 mmol/L)
Potassium	3.8 mEq/L	(3.8 mmol/L)
Bicarbonate	24 mEq/L	(24 mmol/L)
Urea	24 mg/dL	(8.6 mmol/L)
Creatinine	1.2 mg/dL	(109 umol/L)

DIAGNOSIS

- The patient is polycythemic, that is, has a hemoglobin of 17 g/dL (170 g/L) and erythrocyte count of $6.19 \times 10^6/\mu L$ ($6.19 \times 10^{12}/L$). The platelet count is also raised and the patient has splenomegaly. In the first instance, one should consider the diagnosis of primary proliferative polycythemia, also known as polycythemia rubra vera. Secondary and relative causes of polycythemia were excluded. The hyperkalemia seen in the serum sample may be a spurious finding. A repeat collection was performed on a plasma sample

---

WHAT IS YOUR EXPLANATION FOR THE HYPERKALEMIA SEEN IN THE SERUM SAMPLE AND NOT IN THE PLASMA SAMPLE?

- Check the platelet count from the hematology report. A myriad of changes in the platelet size, shape, density, cytoplasm, surface membrane receptors, glycoproteins, coagulant activity, arachidonic acid metabolism, and aggregation have all been catalogued in patients with myeloproliferative disorders. Platelets are rich in potassium, which may be released on clotting. Plasma samples overcome this problem

TABLE
Secondary and relative causes of polycythemia

Due to compensatory erythropoietin increase in:
- High-altitude environments
- Pulmonary disease and alveolar hypoventilation
- Increased-affinity hemoglobins
- Heavy smoking
- Methemoglobinemia (rare)

Due to inappropriate erythropoietin increase in:
- Renal disorders
- Hepatocellular carcinoma
- Massive uterine fibromyomata
- Cerebellar hemangioblastoma

Relative
- "Stress" or "spurious" polycythemia
- Dehydration: water deprivation, vomiting
- Plasma loss: burns, enteropathy

POINTS TO REMEMBER
- Consideration of a thrombocytosis as a cause of spurious hyperkalemia in clotted samples
- Possibility of ACE (such as captopril) inhibitors causing a secondary hypoaldosteronism and hence hyperkalemia
- The laboratory and clinical findings in polycythemia rubra vera
- Alteration of platelet characteristics in patients with myeloproliferative disorders

REFERENCES
1. Messinezy M, Pearson TC. Polycythemia. Mol Aspects Med 1996;17:189–207.
2. Kutti J, Wadenvik H. Diagnostic and differential criteria of essential thrombocythemia and reactive thrombocytosis. Leuk Lymphoma 1996;22:41–5.
3. Murphy S. Polycythemia vera. Dis Mon 1992;38:153–212.
4. Rosenthal DS. Clinical aspects of chronic myeloproliferative diseases. Am J Med Sci 1992;304:109–24.
5. Fields W, Freeman NJ. The hypercoagulability of polycythemia vera. Hosp Pract (Off Ed) 1993;28:65–8.

THIS CASE WAS PRESENTED BY DR. JOHN O'CONNOR, EASTBOURNE HOSPITALS NHS TRUST, EAST SUSSEX, UK

15. Sickness in pregnancy

PRESENTATION

HISTORY OF PRESENTING COMPLAINT
- Day 1 (Friday): A 20-y-old primigravida with α-thalassemia trait was seen at 35 wk gestation at the prenatal clinic. She complained of a 5-d history of diarrhea and vomiting
- Her pregnancy had been uneventful to date apart from one admission with a urinary tract infection

ON EXAMINATION
- Abdominal tenderness, especially in right hypochondrium and right renal angle
- No clinically detectable fetal abnormality

MANAGEMENT
- Arranged for urea and electrolytes and liver function tests and allowed patient to return home

RESULTS

Sodium	140 mEq/L	(140 mmol/L)
Potassium	2.8 mEq/L	(2.8 mmol/L)
Urea	3 mg/dL	(1.2 mmol/L)
Creatinine	0.5 mg/dL	(46 µmol/L)
Alanine aminotransferase	12 IU/L	(12 IU/L)
Total protein	6.7 g/dL	(67 g/L)
Alkaline phosphatase	221 IU/L	(221 IU/L)
Albumin	3.4 g/dL	(34 g/L)
Bilirubin	0.9 mg/dL	(15 µmol/L)
Hemoglobin	12.7 g/dL	(127 g/L)
Platelets	$191 \times 10^3/\mu L$	(191×10^9/L)
Leukocyte count	$9.3 \times 10^3/\mu L$	(9.3×10^9/L)

FURTHER ACTION
- The laboratory was unable to contact anyone at the clinic about the potassium concentration because it was an audit afternoon (staff was unavailable due to participation in a case-review session). The midwife contacted on Monday (Day 4) arranged for an appointment for the next day
- Day 5: At the clinic, the patient's gastrointestinal symptoms were improving, but she was still complaining of tiredness and weakness, especially in the legs
- A repeat potassium concentration was 2.6 mEq/L (2.6 mmol/L)

WHAT COULD BE THE CAUSE OF HER SYMPTOMS?

WHAT FURTHER MANAGEMENT IS INDICATED?

DIFFERENTIAL DIAGNOSIS
- Diarrhea and vomiting: Food poisoning, gastrointestinal infection
- Abdominal pain: Related to above, or urinary tract infection, cholecystitis, or related to pregnancy
- Weakness: Effect of hypokalemia

MANAGEMENT
- Admit, and replenish potassium
- Slow-release potassium tablets prescribed

PROGRESS
- Day 6: Slow-release potassium caused nausea and heartburn; changed to soluble potassium
- Day 7: Hypertension noted. Arranged for renal biochemistry profile. Results as above but aspartate aminotransferase 171 IU/L and alanine aminotransferase 34 IU/L. Obstetric staff reassured
- Day 8: Polydipsia and polyuria noted. (Day 7 input 4530 mL, output 4500 mL)
- Obstetric staff decided to restrict fluids, but was concerned because despite the baby appearing to be doing well, they could not rationalize the mother's symptoms:
  - Polyuria and polydipsia
  - Hypokalemia
  - Weakness and leg pain
  - Transient hypertension

WHAT DIAGNOSIS WILL FIT THESE FINDINGS?

WHAT FURTHER INVESTIGATIONS AND MANAGEMENT ARE INDICATED?

DIFFERENTIAL DIAGNOSIS
- As before but in light of hypertension, possibility of Conn's syndrome and pheochromocytoma also occur

MANAGEMENT
- The obstetric team contacted the clinical chemistry department for advice on diagnosis and management

The patient's history was reassessed taking note of:
- Prolonged diarrhea and vomiting, now stopped
- Hypokalemia
- Several references to weakness and pain in legs confirmed on direct questioning
- She walked everywhere or used public transport
- She was noted to have general weakness but no muscle tenderness

The following diagnoses were considered:
- Hypokalemia due to prolonged diarrhea and vomiting
- Rhabdomyolysis due to hypokalemia
- Tubular dysfunction due to hypokalemia or myoglobinuria

- Reanalysis of previous samples for serum creatine kinase revealed:
  Day 5   9158 IU/L
  Day 6   8516 IU/L
  Day 7   6468 IU/L
  Day 8   3998 IU/L
  Day 9   1418 IU/L

- Urine potassium 13 mEq/L (13 mmol/L), osmolality 51 mOsm/kg (51 mOsm/kg)

TREATMENT
- Recommended continue potassium replacement and allow free fluids to drink

PROGRESS
- Day 11: Potassium 4.1 mEq/L (4.1 mmol/L), creatine kinase 485 IU/L
- Day 13–14: Spontaneous rupture of membranes; safe delivery of baby boy, weight 6 pounds 11 ounces (3.5 kg)
- Day 18: Discharged home. Potassium 3.7 mEq/L (3.7 mmol/L) but stable

FINAL DIAGNOSIS
- This woman had hypokalemia after prolonged vomiting and diarrhea, which also left her with some abdominal tenderness. Exercise undertaken during her daily activities (she walked everywhere) led to rhabdomyolysis. Tubular damage, which can occur in hypokalemia and may have been potentiated by myoglobinuria, was probably responsible for the polyuria and thus the polydipsia. The hypertension was transient and not repeated

PATHOPHYSIOLOGY
- Rhabdomyolysis is a well-described, if little known, complication of hypokalemia. It has been documented in association with a wide variety of causes, including licorice ingestion, diuretic and

- laxative abuse, celiac disease, Bartter's and Conn's syndrome, and even just drinking too much tea!
- The mechanism appears to be due to muscle ischemia. Vasodilation in muscles appears to be triggered by the release of potassium from the contracting muscle. In the presence of hypokalemia, the vasodilation may be insufficient to ensure adequate oxygen delivery to the tissue, and ischemic damage ensues
- If looked for, the phenomenon appears to be common, some evidence for muscle enzyme release being found in 28% of patients with a potassium concentration <3.5 mEq/L (3.5 mmol/L)

POINTS TO REMEMBER
- Rhabdomyolysis is a complication of hypokalemia
- Unexpected electrolyte results need timely notification to clinical teams
- Clinical teams faced with unusual results may need assistance on diagnosis and management
- Severe electrolyte disorders may have consequences in tissues
- Rhabdomyolysis may be caused by hypokalemia (and hypophosphatemia)
- Rhabdomyolysis may cause renal damage

REFERENCES
1. Singhal PC, Venkatesan J, Gibbons N, Gibbons J. Prevalence and predictors of rhabdomyolysis in patients with hypokalemia. New Engl J Med 1990;323:1488.
2. Knochel JP, Sclein EM. On the mechanism of rhabdomyolysis in potassium depletion. J Clin Invest 1972;51:1750–8.
3. Trewby PN, Rutter MD, Earl UM, Sattar MA. Teapot myositis. Lancet 1998;351:1248.
4. Knochel JP. Neuromuscular manifestations of electrolyte disorders. Am J Med 1982;72:521–35.
5. Smith JD, Perazella MA, DeFronzo RA. Hypokalemia. In: Arieff AI, DeFronzo RA, eds. Fluid, electrolyte and acid base disorders. New York: Churchill-Livingstone, 1995:387–426.

THIS CASE WAS PRESENTED BY DR. TREVOR GRAY, NORTHERN GENERAL HOSPITAL, SHEFFIELD, UK

16. Look before you leap

PRESENTATION

HISTORY OF PRESENTING COMPLAINT
- A 61-y-old woman presented to her primary-care physician with depression. He checked her full blood count, thyroid function, urea, and electrolytes. Serum potassium was 2.6 mEq/L (2.6 mmol/L)
- She was advised to eat a high-potassium diet. The serum potassium varied between 2.6 and 3.0 mEq/L (2.6 and 3.0 mmol/L) over the next 2 mo
- She was referred to the hospital for advice

PAST MEDICAL HISTORY
- Diagnosed as hypertensive 4 y previously, with a serum potassium of 3.0 mEq/L (3.0 mmol/L)
- Initially given bendrofluazide, but then blood pressure normalized off treatment
- Left middle cerebral artery cerebrovascular accident 2 y previously; potassium at that time was normal

DRUG HISTORY
- On hormone-replacement therapy and antidepressants

ON EXAMINATION
- Not Cushingoid in appearance
- Blood pressure was 160/100 mmHg

INITIAL INVESTIGATIONS

Serum

Sodium	140 mEq/L	(140 mmol/L)
Potassium	2.9 mEq/L	(2.9 mmol/L)
Bicarbonate	29 mEq/L	(29 mmol/L)
Urea	11 mg/dL	(3.8 mmol/L)
Creatinine	1.0 mg/dL	(84 µmol/L)
Glucose	67 mg/dL	(3.7 mmol/L)

Normal liver function tests, calcium, and thyrotropin results

---

WHAT DIAGNOSES WOULD YOU CONSIDER?

HOW WOULD YOU FURTHER INVESTIGATE HER?

PROVISIONAL DIAGNOSES
- Low intake of potassium or excessive loss (renal or intestinal)
- Conn's syndrome
- Cushing's syndrome

INVESTIGATIONS
- 24-h urine potassium concentrations were requested, and she was taken off potassium supplementation
- On first occasion she had:

Potassium output	34 mEq/24 h	(34 mmol/24 h)
Serum potassium	2.9 mEq/L	(2.9 mmol/L)

  It was noted that urine volume was 5.2 L
- Repeat 24-h collection showed:

Potassium output	36 mEq/24 h	(36 mmol/24 h)
Serum potassium	3.6 mEq/L	(3.6 mmol/L)
24-h urine volume	5.4 L	

- At this stage the patient volunteered that some years previously (when troubled with menopausal symptoms) she had been very thirsty with polyuria and nocturia eight times a night
- Plasma renin and aldosterone concentrations were measured

  Plasma renin activity

Supine	3.8 pmol/h/mL	(4.7 ng/h/mL)
1/2 h ambulant	3.9 pmol/h/mL	(4.8 ng/h/mL)

  Aldosterone

Supine	7.9 ng/dL	(220 pmol/L)
4 h ambulant	9.7 ng/mL	(270 pmol/L)

HOW WOULD YOU INTERPRET THESE RESULTS?

WHAT FURTHER INVESTIGATIONS WOULD YOU REQUEST?

- The urine potassium loss is relatively high
- These plasma renin and aldosterone concentrations are not compatible with a diagnosis of primary hyperaldosteronism provided the patient is not on any medication (nor for several weeks before examination)
- Slightly elevated renin concentrations like these are often seen in patients on diuretics
- Potassium depletion can reduce aldosterone concentrations (so aldosterone can be normal even in Conn's)

FURTHER INVESTIGATIONS
- She was admitted for a water-deprivation test
- When admitted, her serum potassium was 2.9 mEq/L (2.9 mmol/L), though she claimed to be taking Sando K
- She denied diuretic use, but admitted occasional laxatives

Water deprivation test

TIME (h)	SERUM OSMOLALITY (mOsm/kg)	URINE OSMOLALITY (mOsm/kg)
0800	298	–
0815	–	78
0915	–	77
1000	290	71
1100	–	79
1200	289	81
1315	–	77
1400	296	80
1500	–	73
1600	288	61
Desmopressin acetate administered		
1910	–	178
2145	–	354
0415	–	362
0645	–	370

- Initial urine osmolality low, there was no change over next few hours, and there was no change in weight. Serum osmolalities were all normal. The patient was presumed to be finding water and continuing to drink
- There was some response to desmopressin acetate, although not normal
- A diagnosis was suggested of nephrogenic diabetes insipidus as a result of vacuolation of renal tubular cells consequent to prolonged hypokalemia
- Two urine samples during the day's collections were noted to be pink, one slightly, one markedly so. The color was an indicator. The color became more pronounced with age of urine as it became more alkaline
- It was noted that the lady was slim, with a weight of 105 pounds (47.7 kg) and a body-mass index of 20
- Phenolphthalein was identified in a urine sample. It is an ingredient of several laxatives. It is colorless at pH 8.2 and red at pH >10

- The patient admitted only occasional use of senna. She was prescribed potassium supplements; some months later, her serum potassium was 3.4 mEq/L (3.4 mmol/L)
- She developed some angina on exertion, but had a normal exercise test

FINAL DIAGNOSIS: Laxative-induced hypokalemia leading to nephrogenic diabetes insipidus

KEY FEATURES
- Usually, moderate potassium depletion → intracellular acidosis → increased renal H^+ excretion → metabolic alkalosis
  - There may be Cl^- depletion
  - This patient had serum bicarbonate 29 mEq/L (29 mmol/L), chloride 99 mEq/L (99 mmol/L), and potassium 2.9 mEq/L (2.9 mmol/L)
  - In contrast to protracted vomiting, in diarrhea or laxative abuse intestinal losses are rich in potassium bicarbonate, so potassium depletion may develop with a metabolic acidosis
  - This woman's serum bicarbonate was not raised
- Potassium depletion leads to increased ammonia production with a substantial increase in urine pH (which helped the urine samples to appear pink). It causes lysosomal swelling and vacuolation of nucleated cells, and reduces the number of antidiuretic hormone (and parathyroid hormone) receptors on cell surfaces. K^+ depletion can also lead to suppression of cortical aquaporin-2 and downregulation of medullary aquaporin-2. A nephrogenic diabetes insipidus results
- Potassium depletion can also stimulate the thirst mechanism
- Note also that chronic potassium depletion can cause cardiac rhythm abnormalities

POINTS TO REMEMBER
- Use one's eyes and note unusual findings, such as color and 24-h urine volumes
- Patients are sometimes economical with the information they divulge
- Hypokalemia can cause nephrogenic diabetes insipidus

REFERENCES
1. O'Reilly DS. Increased ammoniagenesis and the renal tubular effects of potassium depletion. J Clin Pathol 1984:37:1358–62.
2. Marples D, Frokiaer J, Dorup J, Knepper MA, Nielsen S. Hypokalaemia-induced downregulation of aquaporin-2 water channel expression in rat kidney medulla and cortex. J Clin Invest 1996:97:1960–8.
3. Amlal H, Krane CM, Chen Q, Soleimani M. Early polyuria and urinary concentrating defect in potassium deprivation. Am J Physiol Renal Physiol 2000:279:F655–63.

THIS CASE WAS PRESENTED BY DR. CERIDWEN E. DAWKINS, FRENCHAY HOSPITAL, BRISTOL, UK

17. A 52-y-old man with abdominal pain and vomiting

PRESENTATION

HISTORY OF PRESENTING COMPLAINT
- A 52-y-old man who worked as a decorator was admitted to the emergency department with a 1-d history of abdominal pain and vomiting
- The pain was severe, sharp, and localized to the center of the abdomen. It did not radiate
- The vomiting had been profuse and was now watery. He was barely able to keep fluids down

DIRECT QUESTIONING
- No food since a very heavy bout of drinking at a party 4 d previously
- No alcohol since that time

PAST MEDICAL HISTORY
- Nothing of note. There was no prior history of diabetes mellitus

SOCIAL HISTORY
- He was a known alcoholic [intake ~1 bottle of spirits per day (~250 g/d of ethanol)]
- He smoked ~1 pack of cigarettes a day

FAMILY HISTORY
- No family history of diabetes mellitus

ON EXAMINATION
- He was tachypneic: His respiratory rate was 30/min; his pulse rate was 92 beats/min
- His sclera were jaundiced
- He was moderately dehydrated, but well perfused
- He had a tender central abdomen, but there was no rebound tenderness or guarding
- No other abnormalities were found

WHAT ARE YOUR PROVISIONAL DIAGNOSES?

WHAT INVESTIGATIONS ARE INDICATED IN THIS PATIENT?

PROVISIONAL DIAGNOSES
- Pancreatitis
- Diabetic ketoacidosis
- Alcoholic hepatitis
- Gastritis

INITIAL INVESTIGATIONS
Blood
pH	7.29	(H^+ 51 nmol/L)
pCO_2	11.3 mmHg	(1.5 kPa)
pO_2	126 mmHg	(16.7 kPa)

Serum
Sodium	125 mEq/L	(125 mmol/L)
Potassium	3.2 mEq/L	(3.2 mmol/L)
Bicarbonate	5 mEq/L	(5 mmol/L)
Urea	18 mg/dL	(6.4 mmol/L)
Creatinine	2.9 mg/dL	(256 µmol/L)

Plasma
Glucose	151 mg/dL	(8.4 mmol/L)
Bilirubin	3.6 mg/dL	(62 µmol/L)
γ-Glutamyltransferase	891 IU/L	(891 IU/L)
Aspartate aminotransferase	180 IU/L	(180 IU/L)
Amylase	220 IU/L	(220 IU/L)

WHAT IS YOUR WORKING DIAGNOSIS?

WHAT FURTHER INVESTIGATIONS WOULD YOU REQUEST?

WORKING DIAGNOSES
- Partially compensated metabolic acidosis a possible cause. Note: The pH is low ([H⁺] is high) with a low pCO$_2$
- Alcoholic hepatitis or gastritis

FURTHER INVESTIGATIONS
- The laboratory initiated further investigation because there was no obvious cause of the acidosis (diabetes mellitus and a renal cause could be excluded)
  - Plasma chloride 89 mEq/L (89 mmol/L)
  - Anion gap 35 mEq/L (35 mmol/L). Thus, there is a high anion gap metabolic acidosis
  - Plasma lactate 9 mg/dL (0.99 mmol/L)
- A toxic cause was sought
  - Acetaminophen and aspirin: Not detected
  - Ethanol 18.4 mg/dL (4 mmol/L)
- Because of the history and the disproportionally raised creatinine concentration, plasma ketone concentration was estimated by Ketostix at 58 mg/dL (10 mmol/L)
- Subsequently, β-hydroxybutyrate was measured: 103 mg/dL (10,100 μmol/L)

FINAL DIAGNOSIS: Alcoholic ketoacidosis (AKA)

MANAGEMENT
- He was treated with intravenous fluids (normal saline and 5% and 10% dextrose) analgesics, and antiemetics
- He responded well, and the acidosis improved over the following 2 d. He made an uneventful recovery. He was advised to get counseling for his alcohol intake

KEY FEATURES

PRESENTATION
- Key features are metabolism of alcohol, starvation, vomiting, and the time course of events leading to admission
- Dehydration, tachypnea, and possible jaundice on admission
- Differential diagnosis includes gastritis, alcoholic hepatitis, and pancreatitis (which may accompany the acidosis)

BIOCHEMISTRY (SEE FIGURE)
- Hepatic alcohol metabolism increases the ratio of the reduced form of nicotinamide-adenine dinucleotide to the oxidized form (NADH-NAD⁺ ratio)
- Glucose production falls due to depletion of glycogen stores by starvation and the reduction of hepatic gluconeogenesis due to the raised NADH-NAD⁺ ratio
- Raised concentrations of cortisol and catecholamines increase lipolysis
- Free fatty acids are converted to ketones and the raised NADH-NAD⁺ ratio favors the production of β-hydroxybutyrate

DIAGNOSIS
- Plasma electrolyte concentrations are not characteristic, and plasma glucose concentration is variable
- Plasma alcohol is often undetectable, and plasma lactate is not usually sufficiently raised to explain the degree of acidosis

- Plasma ketones, assessed by Ketostix, may be negative or mildly positive. The source of the ketoacidosis is mainly due to an increased concentration of β-hydroxybutyrate, which is not detected by the nitroprusside Ketostix reaction
- An indirect indicator of the ketosis is the concentration of creatinine (measured by the Jaffé reaction), which is often disproportionally raised because of interference by ketones with the reaction (also seen in diabetic ketoacidosis)

TREATMENT
- Treatment with insulin and bicarbonate is unnecessary, and intravenous glucose in combination with saline allows quicker recovery
- Potassium supplementation is often necessary, and plasma phosphate concentration should be monitored and treatment given if indicated

FIGURE   THE PATHOPHYSIOLOGY OF ALCOHOLIC KETOACIDOSIS

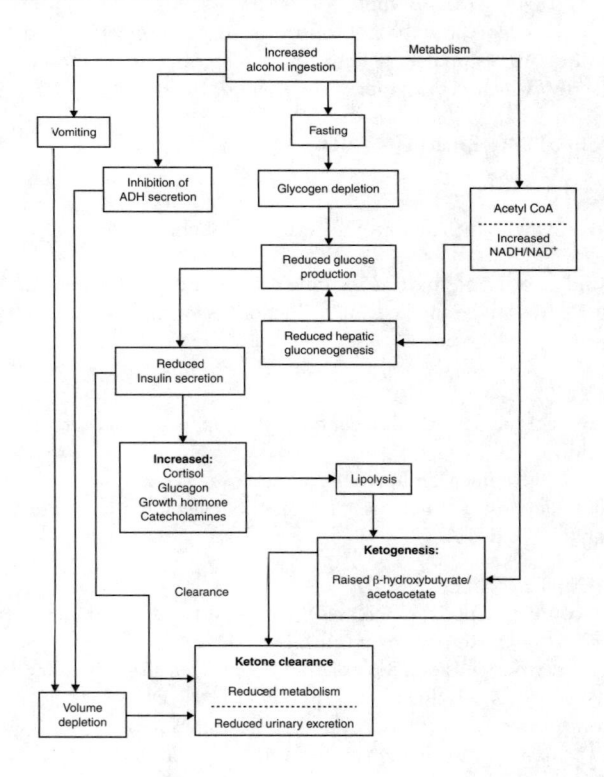

Figure reprinted with permission from Hooper RJL. Alcoholic ketoacidosis: the late presentation of acidosis in an alcoholic. Ann Clin Biochem 1994;31:579–82.

POINTS TO REMEMBER
- Know how to diagnose metabolic acidosis
- Recognize the presenting features of AKA
- Recognize the biochemical features of AKA
  - The metabolic acidosis
  - The disproportionally raised creatinine relative to urea (due to the interference of ketones with creatinine)
- Know the differential diagnosis of AKA (alcoholic hepatitis, pancreatitis, diabetic ketoacidosis, and poisoning)
- Understand the appropriate treatment of AKA with fluids and glucose

REFERENCES
1. Thompson CJ, Johnston DG, Baylis PH, Anderson J. Alcoholic ketoacidosis—an underdiagnosed condition? Brit Med J 1986;292:463–5.
2. Isselbacher KJ. Metabolic and hepatic effects of alcohol. N Engl J Med 1977;296:612–6.
3. Pannall PR. The clinical biochemistry of the alcohol abuser. Clin Biochem Rev 1989;10:158–65.
4. Hooper RJL. Alcoholic ketoacidosis: the late presentation of acidosis in an alcoholic. Ann Clin Biochem 1994;31:579–82.

THIS CASE WAS PRESENTED BY DR. JAMES HOOPER, THE ROYAL BROMPTON HOSPITAL, LONDON, UK

18. An A to Z of intoxication

PRESENTATION

HISTORY OF PRESENTING COMPLAINT
- A 40-y-old woman was admitted comatose to the accident and emergency department after being found semiconscious in bed at home. She was alleged to have taken an overdose of zopiclone tablets (a hypnotic acting on the benzodiazepine receptor) with a large quantity of alcohol

ON EXAMINATION
- She was breathing spontaneously, but unresponsive to stimuli
- Pulse was 50/min (electrocardiogram sinus rhythm), blood pressure 118/70 mmHg, respiration 18/min, and temperature (axilla) 32 °C
- Her pupils were dilated and not reactive to light

INITIAL INVESTIGATIONS

Serum

Sodium	147 mEq/L	(147 mmol/L)
Potassium	4.4 mEq/L	(4.4 mmol/L)
Chloride	103 mEq/L	(103 mmol/L)
Urea	4.0 mg/dL	(1.5 mmol/L)
Creatinine	1.0 mg/dL	(91 µmol/L)
Glucose	305 mg/dL	(16.9 mmol/L)
Amylase	111 IU/L	(111 IU/L)
Alkaline phosphatase	125 IU/L	(125 IU/L)
Alanine aminotransferase	43 IU/L	(43 IU/L)
γ-Glutamyltransferase	352 IU/L	(352 IU/L)
Bilirubin	0.4 mg/dL	(6 µmol/L)

Blood

pH	6.73	(H^+ = 234 nmol/L)
pCO_2	47 mmHg	(6.2 kPa)
pO_2	385 mmHg	(51.2 kPa) (on oxygen)
Bicarbonate	6 mEq/L	(6 mmol/L)

SUGGEST POSSIBLE DIFFERENTIAL DIAGNOSES.

WHAT FURTHER INVESTIGATIONS OR CALCULATED PARAMETERS WILL HELP TO ELUCIDATE THE DIAGNOSIS?

- The patient has an extremely severe metabolic acidosis with a raised anion gap and a normal plasma glucose.

$$\begin{aligned}\text{Anion gap} &= (Na + K) - (Cl + HCO_3) \\ &= (147 + 4) - (103 + 6) \\ &= 42\,mEq/L\ (42\ mmol/L)\end{aligned}$$

DIFFERENTIAL DIAGNOSIS
- Lactic acidosis
- Poisoning:
  - Salicylate
  - Acetaminophen
  - Theophylline
  - Methanol or ethylene glycol
- Alcoholic ketoacidosis
- Organic acidoses [such as medium-chain acyl coenzyme A dehydrogenase deficiency (MCAD)]

FURTHER INVESTIGATIONS

Acetaminophen	<1.5 mg/dL	(<0.1 mmol/L)
Salicylate	<1.0 mg/dL	(<0.1 mmol/L)
Osmolality	470 mOsm/kg	(470 mOsm/kg)

HOW DOES THIS INFORMATION REFINE YOUR DIAGNOSIS?

WHAT ADVICE WOULD YOU GIVE TO THE CLINICIAN?

- The calculated plasma osmolality is 319 mOsm/kg (319 mOsm/kg), using the simplest formula [2(Na+K) + urea + glucose] (for a discussion of calculation formulae, see reference 1)
- This result means the patient has an osmolar gap (difference between measured and calculated osmolarity) of 151 mOsm/kg (151 mOsm/kg)
- An increased osmolar gap is caused by the presence of unmeasured osmoles, notably alcohols (ethanol, methanol, ethylene glycol), mannitol, and glycine (sometimes absorbed from irrigation fluid used during operative procedures, such as transurethral resection of the prostate)

WORKING DIAGNOSIS

- The combination of the metabolic acidosis with a raised anion gap and a raised osmolar gap is strongly suggestive of poisoning with methanol or ethylene glycol. The patient's raised γ-glutamyltransferase also suggests a history of alcohol abuse
- Confirmation of this diagnosis with urgent assays of methanol or ethylene glycol is required, but if these assays are not immediately available, the evidence above is sufficient to begin treatment immediately pending confirmation of the diagnosis. However, it should be noted that a low osmolar gap [<20 mOsm/kg (<20 mOsm/kg)] does not exclude methanol or ethylene glycol poisoning, especially if presentation is late (2)
- Ethylene glycol contributes less to the osmolar gap than methanol, because of its higher molecular weight

KEY FEATURES

- Methanol poisoning is caused by the metabolism of methanol to formic acid, which is a metabolic dead end (see figure) and therefore accumulates, causing the acidosis. Because the systemic toxicity is caused by a metabolite, there is normally a latent period (12–72 h after ingestion) before symptoms arise
- Patients present with headache, weakness, sweating, nausea, and vomiting. There is a severe metabolic acidosis and bradycardia, and respiratory failure may develop. Conscious patients frequently have visual symptoms, ranging from diminished visual acuity, photophobia, and visual field defects to complete blindness

MANAGEMENT

- Because the toxic metabolite (formic acid) is formed by metabolism of methanol, its production can be halted by inhibition of alcohol dehydrogenase. A specific alcohol dehydrogenase inhibitor (4-methyl pyrazole; fomepizole) is now available, and treatment with fomepizole is highly effective in preventing further formation of metabolite (3). However, it is expensive and not yet available in many countries. The classic alternative is to saturate the alcohol dehydrogenase system by giving ethanol (orally for minor intoxication, intravenously in severe cases) to produce a plasma ethanol concentration of ~100–200 mg/dL (~21–42 mmol/L). Methanol is metabolized at a rate ~15% that of ethanol. Therapy should be continued until plasma methanol concentrations are <20 mg/dL (<6.2 mmol/L). Hemodialysis may also be required in severe cases (4)
- Aggressive correction of acidosis with intravenous sodium bicarbonate is essential. The mortality is ~20% if the plasma bicarbonate concentration is >20 mEq/L (20 mmol/L) on admission, rising to 50% if severe acidosis is present on admission [plasma bicarbonate <10 mEq/L (10 mmol/L)]. As well as formic acid, lactic acid also contributes to the acidosis, formed due to depletion of nicotinamide-adenine dinucleotide by oxidation of alcohols (see figure)

FIGURE   METABOLISM OF METHANOL

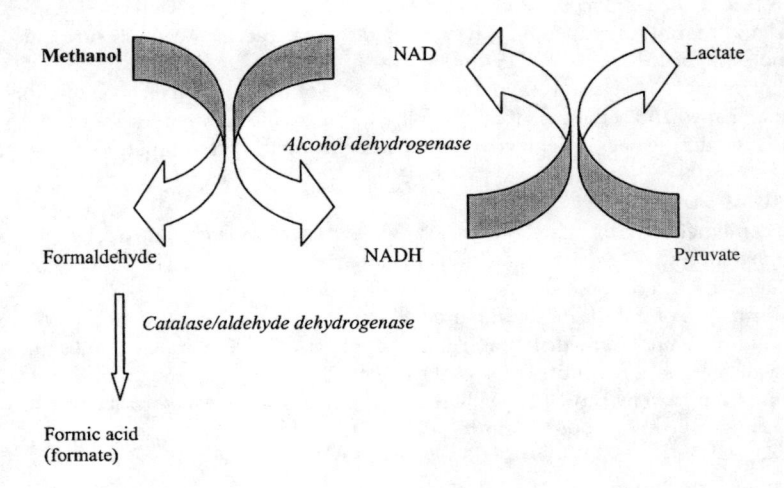

Methanol → Formaldehyde (Alcohol dehydrogenase; NAD → NADH; Pyruvate → Lactate)

Formaldehyde → Formic acid (formate) (Catalase/aldehyde dehydrogenase)

FURTHER INFORMATION
- Methanol is used as an industrial raw material and solvent and is frequently found in paint thinners and car windshield wiper fluids, which can be almost entirely methanol. Methylated spirit is normally ~5% methanol and rarely causes problems when ingested, because the large ethanol component overwhelms the methanol, and metabolism to toxic intermediates is largely prevented. In general, patients who ingest methanol-ethanol mixtures (for example, as contaminated spirits) do better than those who ingest methanol alone, because the excess of ethanol is protective to some extent. As little as 30 mL of pure methanol can be fatal (less in children)

FINAL DIAGNOSIS: Methanol poisoning
- Methanol ingestion was confirmed in this patient; plasma methanol concentration some 30 h after ingestion was 63 mg/dL (19.7 mmol/L). She had also consumed significant amounts of ethanol [plasma ethanol concentration at the same time was still 46 mg/dL (10.0 mmol/L)]. Despite treatment with bicarbonate and intravenous ethanol, she died 5 d after admission. Subsequent investigation established that she had purchased illicit spirits to which methanol had been added to increase the alcoholic strength. A substantial amount of this material had been ingested 20–24 h before admission to the hospital

POINTS TO REMEMBER
- Understand the use and limitations of the anion gap and osmolar gap in the differential diagnosis of metabolic acidosis
- Recognize the mode of presentation and treatment of methanol poisoning
- Understand the biochemical basis of methanol poisoning

REFERENCES

1. Kruse JA, Cadnapaphornchai P. The serum osmole gap. J Crit Care 1994;9:185–97.
2. Glaser DS. Utility of the serum osmol gap in the diagnosis of methanol or ethylene glycol ingestion. Ann Emerg Med 1996;27:343–6.
3. Jacobsen D, Martin KE. Antidotes for methanol and ethylene glycol poisoning. J Toxicol Clin Toxicol 1997;35:127–43.
4. Kruse JA. Methanol poisoning. Intensive Care Med 1992;18:391–7.

THIS CASE WAS PRESENTED BY MR. MICHAEL HALLWORTH, ROYAL SHREWSBURY HOSPITAL, SHREWSBURY, UK

19. Cushing's syndrome: one patient, two causes?

PRESENTATION

HISTORY OF PRESENTING COMPLAINT
- A 56-y-old woman was referred to the endocrine outpatient clinic
- She had visited her primary-care physician's office 2 mo previously complaining of increasing weight, easy bruising of the skin, and difficulty walking up stairs
- Her primary-care physician thought she had the typical features of Cushing's syndrome and arranged a 24-h urine free cortisol excretion measurement, which was elevated at 362 µg/24 h (1001 nmol/24 h)

DIRECT QUESTIONING
- She reported an increase in facial hair, which she had started to shave on alternate days

PAST MEDICAL HISTORY
- Nonsmoker, with moderate intake of alcohol (14 units per week)
- Nothing else of note

SOCIAL HISTORY
- Nothing of note

FAMILY HISTORY
- Nothing of note

DRUG HISTORY
- She was not taking any drugs

ON EXAMINATION
- She had marked facial plethora with a degree of moon facies, thin skin with numerous bruises, truncal adiposity, and a proximal myopathy, which was quite marked in the lower limbs
- Her cardiovascular system was normal apart from a consistent finding of hypertension ranging from 170/90 to 180/120 mmHg. She had no neuro-ophthalmic signs
- She was admitted to the endocrine ward for further investigations

---

WHAT INVESTIGATIONS WOULD YOU REQUEST?

INVESTIGATIONS

- Two midnight sleeping serum cortisols were 6.3 and 5.8 µg/dL (176 and 163 nmol/L)
- Her baseline 0900 h serum cortisol was 25.9 µg/dL (726 nmol/L), which suppressed to 16.0 µg/dL (442 nmol/L) after 48 h of dexamethasone 0.5 mg four times a day (qds) and then suppressed further to 10.7 µg/dL (301 nmol/L) after 48 h of dexamethasone 2 mg qds
- Two 0900 h plasma adrenocorticotropic hormone (ACTH) measurements were 103 and 97 pg/mL (103 and 97 ng/L)
- Serum K^+ was 4.1 mEq/L (4.1 mmol/L)
- Computed tomography (CT) and magnetic resonance imaging scans of the pituitary were reported as normal
- CT of the adrenals showed both glands were hyperplastic with no apparent focal abnormality
- Chest X-ray was normal

WORKING DIAGNOSES

ACTH-dependent Cushing's syndrome arising from one of the following:
- Pituitary ACTH overproduction [Cushing's disease (CD)]
- Ectopic ACTH secretion (EAS)

Because the patient does not have hypokalemia (a feature most closely associated with EAS) *(1)* or an obvious lesion for the source of ACTH secretion on routine radiological investigation, the differential diagnosis is between CD and occult EAS (oEAS).

FURTHER INVESTIGATIONS

Simultaneous bilateral inferior petrosal sinus (IPS) catheterization and sampling (IPSCS) with ACTH-releasing hormone (CRH) stimulation has been reported to approach 90–100% sensitivity and specificity in distinguishing patients with CD from patients with oEAS *(2,3)*.

TABLE
Results of inferior petrosal sinus catheterization and sampling

	TIME (MIN)	PERIPHERAL VEIN	PETROSAL VEINS LEFT	PETROSAL VEINS RIGHT
100 µg CRH →	0	57	560	515
	5	74	2760	2520
	10	111	3050	2890
	15	104	2450	2320

ACTH [pg/mL (ng/L)]

DIAGNOSIS: ACTH-dependent Cushing's syndrome due to pituitary ACTH overproduction

MANAGEMENT
- She proceeded to pituitary surgery after 6 wk of metyrapone treatment to lower her circulating cortisol concentrations
- At surgery, a microadenoma was removed. Histology and immunocytochemistry confirmed an ACTH-secreting adenoma
- Five days postsurgery, and 12 h after the last dose of hydrocortisone, her 0900 serum cortisol was 14.7 µg/dL (411 nmol/L) and the patient returned to surgery for a total hypophysectomy. No adenomatous tissue was found on histology

FURTHER INVESTIGATIONS—Postsurgery
- Two months after surgery, the patient still had an elevated midnight sleeping serum cortisol of 9.6 µg/dL (270 nmol/L) and a lack of suppression of 0900 cortisol [2.8 µg/dL (79 nmol/L)] to an overnight dexamethasone (1 mg) suppression test. Her 0900 plasma ACTH concentration was undetectable at <10 pg/mL (<10 ng/L)
- The patient underwent selective adrenal vein sampling and a selenocholesterol scan. Both investigations suggested an area of persistent high cortisol activity in the left adrenal
- She went on to have a unilateral left adrenalectomy, and a 1.5-cm nodule was found in the left adrenal gland. The non-nodular cortex was hyperplastic. Three months postsurgery, the patient was free of the biochemical and clinical features of her Cushing's disease

FINAL DIAGNOSIS: ACTH-dependent Cushing's syndrome due to pituitary ACTH overproduction with subsequent nodular adrenal hyperplasia
- Assessment of this patient after surgery failed to demonstrate cure (4). Subsequent investigations showed that she had a rare feature of ACTH-dependent Cushing's syndrome—unilateral nodular adrenal hyperplasia
- Diffuse bilateral hyperplasia or micronodular hyperplasia is a feature of 70–80% of patients with pituitary-dependent Cushing's disease. About 20–30% have bilateral nodular hyperplasia (nodule size >0.5–5.3 cm) (5), in which one or more nodules coexist with a hyperplastic adrenal cortex. These nodules may be autonomous in function but arise initially as a result of ACTH drive. Patients with nodular hyperplasia give a longer history, and are older and more hypertensive than those with Cushing's disease and diffuse hyperplasia (6). The data suggest that nodular hyperplasia may be a result of long-standing Cushing's disease with varying degrees of ACTH dependence and adrenocortical autonomy

POINTS TO REMEMBER
- Pituitary-dependent Cushing's disease may not always be cured by successful removal of pituitary adenomatous tissue
- The presence of adrenocortical nodular hyperplasia may lead to confusing biochemical and

radiological findings

REFERENCES

1. Orth DN. Cushing's syndrome. New Engl J Med 1995;332:791–801.
2. Oldfield E, Doppman JL, Nieman L, Chrousos GP, Miller DL, Katz DA. Petrosal sinus sampling with and without corticotropin releasing hormone for the differential diagnosis of Cushing's syndrome. New Engl J Med 1991;325:897–905.
3. Kaltsas GA, Giannulis MG, Newell-Price JDC, et al. A critical analysis of the value of simultaneous inferior petrosal sinus sampling in Cushing's disease and the occult ectopic adrenocorticotropin syndrome. J Clin Endocrinol Metab 1999;84:487–92.
4. McCance DR, Besser M, Atkinson AB. Assessment of cure after transphenoidal surgery for Cushing's disease. Clin Endocrinol 1996;44:1–6.
5. Smals AGH, Pieters GF, van Haelst UJG, Kloppenborg PW. Macronodular adrenocortical hyperplasia in longstanding Cushing's disease. J Clin Endocrinol Metab 1984;58:25–31.
6. Hermus AR, Pieters GF, Smals AG, et al. Transition from pituitary-dependent to adrenal dependent Cushing's syndrome. New Engl J Med 1988;318:966–70.

THIS CASE WAS PRESENTED BY PROFESSOR JACKY M. BURRIN AND PROFESSOR JOHN P. MONSON, ST BARTHOLOMEW'S AND ROYAL LONDON SCHOOL OF MEDICINE AND DENTISTRY, LONDON

20. An unusual CK-MB result

PRESENTATION

HISTORY OF PRESENTING COMPLAINT
- A 64-y-old Caucasian man had been receiving treatment for hypercholesterolemia with the 3-hydroxy-3-methylglutaryl coenzyme A (HMG-CoA) reductase inhibitor simvastatin for 3 wk
- Serum creatine kinase (CK) was measured as part of the routine follow-up of statin treatment

PAST MEDICAL HISTORY
- The patient had suffered from angina for 12 y. He had hypertension and epilepsy
- He walked with the aid of crutches due to weakness in his left leg and had a neuropathic bladder. Both of these conditions had been attributed to a corda equina, which had been operated on with a lumbar laminectomy 3 y previously

DRUG HISTORY
- In addition to simvastatin, he received enalapril and sodium valproate. Prophylactic trimethoprim was also given to guard against urinary tract infections. He received no specific treatment for his angina

SOCIAL HISTORY
- Retired concrete block manufacturer

INITIAL INVESTIGATIONS
- Serum CK activity was increased (396 IU/L), of which the cardiac (CK-MB) isoenzyme was 9.6% [38 IU/L, immunoinhibition assay; Vitros Ektachem (Johnson & Johnson, Raritan, NJ) 700 analyzer]

INFORMATION FROM SUBSEQUENT QUESTIONING
- He had no chest pain and denied any recent pain similar to his normal anginal pain

---

WHAT ADVICE WOULD YOU GIVE TO THE PATIENT'S PHYSICIAN REGARDING THE INTERPRETATION OF THIS RESULT?

HOW WOULD YOU EXCLUDE A SPURIOUS CAUSE OF HIS RAISED CK-MB?

- Increases in CK-MB are usually associated with ischemic myocardial damage. Using this particular assay, a CK-MB activity >16 IU/L **and** >4% of the total CK activity is deemed consistent with myocardial infarction. Although the patient had multiple risk factors for an ischemic cardiac event, on this occasion there was no typical history of such. Statin treatment is associated with increases in CK, but this is attributed to the skeletal (MM) isoenzyme only *(1)*. A repeat sample was requested. In the meantime, the patient's physician was advised to carry out an electrocardiogram (ECG) and to stop the simvastatin treatment. Further laboratory studies were undertaken to exclude a spurious analytical cause of his raised CK-MB and to investigate the possibility of a nonmyocardial cause of his raised cardiac isoenzyme fraction
- Immunoinhibition assays use an anti-CK-M serum to inhibit the M subunits of CK-MM and CK-MB. The residual B activity is then assumed to be totally attributable to the B subunit of CK-MB. This assumption is valid for the vast majority of patients, although increased CK-BB activity may be seen in some pathological conditions including certain lung tumors. However, high-molecular-weight forms of CK (macro-CK) occur that may cause diagnostic confusion. Macro-CK type 1 (CK-BB bound to immunoglobulin G, prevalence 0.8–2.3%) and macro-CK type 2 (oligomeric mitochondrial CK, prevalence 0.5–2.6%) may both spuriously elevate CK-MB activity measured by immunoinhibition methods *(2)*
- Several approaches were used to exclude possible spurious analytical elevation of CK-MB. CK-MB sandwich ("mass") immunoassay (CK-MB 23 µg/L, normal <7; Hybritech Icon, San Diego) and electrophoresis (CK-MB 8.5%) results both confirmed the initial immunoinhibition CK-MB assay results. Conversely, troponin I (result not available at initial consultation) was normal (<0.1 µg/L, Beckman Coulter Access, High Wycombe, UK), suggesting the patient had not suffered a myocardial event

---

WHICH NONMYOCARDIAL CONDITIONS MAY BE ASSOCIATED WITH INCREASED CK-MB, AND WHICH TESTS COULD BE USED TO INVESTIGATE THEM?

- True increases in CK-MB have been observed in several other nonmyocardial myopathic situations (3). In this patient there was no evidence of myopathy secondary to renal dysfunction [serum urea 16 mg/dL (5.8 mmol/L), serum creatinine 1.0 mg/dL (90 µmol/L)] or thyroid disease (thyrotropin 1.8 mU/L). There was no history of diabetes mellitus, alcohol abuse (γ-glutamyltransferase 16 IU/L), or rheumatoid arthritis. Serum myoglobin (70 mg/L) was within the reference range. There was no immunological evidence of an inflammatory myopathy (anti-nuclear, –smooth muscle, -mitochondrial, –double-stranded DNA, -Ro, -La, -RNP, -Scl-70, and –Jo-1 antibodies all negative). The patient was not an athlete (4) and had taken no regular strenuous exercise. In many of these situations, the increase in CK-MB has been attributed to a high proportion of regenerating skeletal muscle fibers, which, like fetal skeletal muscle, are relatively rich in CK-MB (5)

FINAL DIAGNOSIS

- Possibly raised CK-MB related to leg weakness
- Possibly raised CK-MB secondary to statin treatment
- This patient presented a diagnostic difficulty in that the original biochemical analysis suggested that he may have suffered an ischemic cardiac event, for which he had multiple risk factors. An ECG, however, showed no evidence of myocardial infarction and the subsequent failure to demonstrate increased troponin I and a "rise and fall" pattern of traditional cardiac markers provided strong evidence against the diagnosis of myocardial infarction. Statin-related increases in CK have, to date, been attributed solely to the MM isoenzyme. In the present case the persistence of the observed increase in serum CK activity and of the MB fraction over the ensuing months after statin withdrawal initially argued against a drug-induced effect (table). It was felt likely that his raised CK-MB related to muscular disease as a result of his left leg weakness. However, the total CK did eventually normalize, raising the possibility that this was a hitherto-unreported side effect of statin treatment

TABLE
Serum cardiac markers at presentation and during follow-up

MARKER	DAY 0	DAY 4[a]	DAY 29	DAY 120	DAY 240
CK (IU/L)	396	530	407	292	138
CK–MB [IU/L (%)]	38 (9.6)	37 (7.0)	28 (7.0)	29 (10.7)	–
CK–MB (µg/L)	23	25	14	25	–
AST (IU/L)	20	24	17	–	–
LDH (IU/L)	639	570	580	–	–
Troponin I (µg/L)	<0.1	<0.1	–	–	–

AST, aspartate transaminase; LDH, lactate dehydrogenase
[a] Electrophoresis on this sample demonstrated 8.5% CK-MB

POINTS TO REMEMBER
- Clinical biochemists must be aware of the many nonmyocardial causes of an increased CK-MB in addition to the problems of laboratory artifact
- Troponin I is an extremely specific marker of myocardial damage, measurement of which is gradually replacing CK-MB

REFERENCES
1. Walker JF. Simvastatin: the clinical profile. Am J Med 1989;87[Suppl 4A]:44S–6S.
2. Moss DW, Henderson AR. Clinical enzymology. In: Burtis CA, Ashwood ER, eds. Tietz textbook of clinical chemistry, 3rd ed. Philadelphia: WB Saunders, 1999:617–721.
3. Chan KM, Ladenson JH. Increased creatine kinase MB in the absence of acute myocardial infarction. Clin Chem 1986;32:2044–51.
4. Thompson GR. Hazards of running a marathon. Br M J 1997;314:1023–5.
5. Siegal AJ, Silverman LM, Evans WJ. Elevated skeletal muscle creatine kinase MB isoenzyme levels in marathon runners. JAMA 1983;250:2835–7.

THIS CASE WAS PRESENTED BY DR. EDMUND J. LAMB, KENT AND CANTERBURY HOSPITALS, CANTERBURY, KENT, AND BY DR. CHRIS J. HEALY, EASTRY PRACTICE, EASTRY, KENT, UK

21. Prolonged jaundice in a neonate

PRESENTATION

HISTORY OF PRESENTING COMPLAINT
- An 8-wk-old infant (corrected age 5 wk) was referred for assessment of prolonged jaundice
- Born at 37/40 after a normal pregnancy
- Well at birth, with no evidence of jaundice
- Discharged at 1 d old, breast fed
- Noted to be jaundiced on day 2
- Day 5: Admitted to local hospital; total bilirubin concentration 25.0 mg/dL (428 µmol/L)
- Received ultraviolet light therapy
- Bilirubin fractions measured:

AGE	TOTAL BILIRUBIN [mg/dL (µmol/L)]	CONJUGATED BILIRUBIN [mg/dL (µmol/L)]
5/7	21.1 (361)	4.3 (73)
8/7	14.7 (251)	4.1 (70)
Patient discharged for outpatient review		
5/52	7.2 (124)	4.1 (70)
6.5/52	7.2 (124)	4.7 (81)

- Thyroid function was normal for a 5-wk-old infant
- Stools noted to be lightly pigmented
- Urine not strongly colored

FAMILY HISTORY
- First child of nonconsanguineous parents
- No family history of liver disease

ON EXAMINATION
- Mild jaundice
- Soft liver edge
- Alert and active

WHAT IS YOUR PROVISIONAL DIAGNOSIS?

WHAT INVESTIGATIONS WOULD YOU REQUEST?

PROVISIONAL DIAGNOSIS
- Conjugated hyperbilirubinemia due to:
  - Biliary abnormalities
  - Infection
  - Inherited metabolic disorders
  - Iatrogenic causes (such as total parenteral nutrition)

INITIAL INVESTIGATIONS

Serum

Total bilirubin	6.9 mg/dL	(118 µmol/L)
Unconjugated (indirect) bilirubin	3.1 mg/dL	(53 µmol/L)
Alkaline phosphatase	1316 IU/L	(1316 IU/L)
Alanine aminotransferase	52 IU/L	(52 IU/L)
Aspartate aminotransferase	92 IU/L	(92 IU/L)
γ-Glutamyltransferase	514 IU/L	(514 IU/L)
Total protein	5.5 g/dL	(55 g/L)
Cholesterol	151 mg/dL	(3.9 mmol/L)
Triglyceride	108 mg/dL	(1.22 mmol/L)
α_1-Antitrypsin concentration	0.126 g/dL	(1.26 g/L)
Phenotype	M	
Amino acids	Negative	
Free thyroxine	1.34 ng/dL	(17.3 pmol/L)
Thyroid-stimulating hormone	6.68 µU/mL	(6.68 mU/L)
Ferritin	17.7 µg/dL	(177 µg/L)
Prothombin time	11 s	
TORCH[a], hepatitis screen	Negative	
Erythrocyte galactose-6-phosphate uridyl transferase	Negative	
Immunoreactive trypsin (IRT)	1.7 µg/dL	(17 µg/L)

Urine

Reducing substances	Negative
Organic acids	Negative
Culture	Negative

Feces

Ultrasound of liver	Small gall bladder
Examination of stools	Not perfectly acholic (slightly pigmented)

[a] TORCH stands for toxoplasmosis, rubella, cytomegalovirus, and herpes simplex viruses

---

WHAT IS YOUR PROVISIONAL OR FINAL DIAGNOSIS?

WHAT FURTHER INVESTIGATIONS WOULD YOU REQUEST, IF ANY?

PROVISIONAL DIAGNOSES
- Biliary abnormality
- Metabolic disorder

FURTHER INVESTIGATIONS
- Technetium trimethyl-1-bromo iminodiacetic acid (TEBIDA) scan (see later for details): No excretion after 24 h
- Liver biopsy: A biopsy showed variable mild to moderate expansion of portal tracts by proliferating bile ductules, fibrosis, and inflammation. The parenchyma showed severe cholestasis, but not steatosis
- Laparotomy and perioperative cholangiogram

FINAL DIAGNOSIS: Extrahepatic biliary atresia

MANAGEMENT
- The infant underwent surgery and received both drug and nutritional support, as detailed below:
  - Surgery: Kasai hepatoportoenterostomy
  - Drug therapy
    - Low-dose oral antibiotic: Ciprofloxacin (25 mg twice a day for 7 d)
    - Bile acid supplement: Ursodeoxycholic acid (20 mg/kg)
  - Nutritional support: Fat-soluble vitamins A, D, E, and K
- She was admitted to hospital on two further occasions with jaundice and evidence of increasingly severe hepatocellular and canalicular cholestasis. She was found to have proliferating bile ductules and inspissated (thickened) bile. These features were those of biliary cirrhosis secondary to extrahepatic biliary atresia and were consistent with a failed Kasai operation. She underwent a liver transplant at 7.5 mo of age

KEY FEATURES
- Biliary atresia is a cholangiopathic panductule disease. In most affected infants, it is present at birth
- Destruction of the extrahepatic and intrahepatic bile ducts leads to cholestasis, fibrosis, and cirrhosis
- There are three subtypes of biliary atresia based on the degree of damage at diagnosis. Type 3, in which the entire extrahepatic biliary tree is involved, is the most common (~85% of cases). Types 1 and 2, in which there is more localized damage, are less common (~10% and extremely rare, respectively)

EPIDEMIOLOGY
- Occurs worldwide in 1/10,000–1/15,000 live births
- Represents a significant cause of neonatal liver disease and is a most important differential diagnosis because it is the main indication for liver transplantation
- Etiology is unclear
- Rarely is there a family history of liver or biliary disease

PRESENTATION
- Birth weight and gestational age is usually normal
- Jaundice (due to conjugated bilirubin) appears shortly after birth and is concomitant with physiological jaundice

- Usually yellow to dark urine with increasingly pale stools, which eventually become acholic. (There may initially be variation in stool color)
- Hepatomegaly is an early feature, and splenomegaly (a late feature) implies hepatic fibrosis
- Weight gain is slow despite adequate feeding
- Atrial and ventricular septal defects, situs invertus, and polysplenia syndrome are associated abnormalities

BIOCHEMISTRY
- Elevated plasma conjugated bilirubin at presentation may range from 2.3 to 11.7 µg/dL (40 to 200 µmol/L). The conjugated bilirubin to total bilirubin ratio is always >0.20
- Plasma aminotransferases are usually elevated, typically in the range of 80 to 200 IU/L; it may be difficult to distinguish from neonatal hepatitis
- Alkaline phosphatase and γ-glutamyltransferase are usually elevated
- Plasma glucose, albumin, and triglycerides are usually normal, but cholesterol may be increased

DIAGNOSIS
- Based on:
  - An absent or contracted gallbladder (after a 4-h fast) on abdominal ultrasound
  - A failure to demonstrate excretion of the radioisotope TEBIDA from the liver into the bowel over a 24-h period after administration, following pretreatment with phenobarbitone (5 mg/kg) for 3–5 d
  - Liver histology that shows features of bile duct obstruction with bile ductule proliferation and portal tract edema; varying degrees of fibrosis and cholestasis
  - A perioperative cholangiogram that may be necessary in ambiguous cases

MANAGEMENT
- Surgery: Kasai hepatoportoenterostomy. The extrahepatic biliary tree is resected completely and biliary continuity restored by anastomosing a loop of jejunum to the transected porta hepatis (the fissure through which the portal vein, hepatic artery, and bile ducts pass on the undersurface of the liver). The earlier the operation is performed (<8 wk of age), the greater the chance of establishing bile flow
- Medical management: Low-dose oral antibiotic to prevent cholangitis
- Nutritional support: Sufficient calories to prevent malnutrition and to overcome fat malabsorption. Oral supplements of fat-soluble vitamins (A, D, E, and K)

PROGNOSIS
- Complications may include recurrent ascending cholangitis, progressive biliary cirrhosis, and portal hypertension. Malnutrition secondary to malabsorption may occur
- Progression to cirrhosis and portal hypertension is inevitable
- If surgery is unsuccessful (failure to re-establish biliary drainage), liver transplantation will be necessary within 1 y
- Children need regular follow up at a specialist center to monitor growth and development, prevent complications, and assess the need for liver transplantation

POINTS TO REMEMBER
- Investigate a neonate with prolonged jaundice promptly
- Measure the bilirubin fractions to determine the type of hyperbilirubinemia present
- Exclude, where possible, other causes of conjugated hyperbilirubinemia

- Be aware that the Kasai operation for biliary atresia is less likely to be successful after the infant is 8 wk old

REFERENCES

1. Kelly D, Green A. Investigation of paediatric liver disease. J Inher Metab Dis 1991;14:531–7.
2. Green A, Morgan I. The term infant: clinical and biochemical problems. In: Neonatology and clinical biochemistry. Cambridge, UK: ACB Venture Publications, 1993: 23–61.
3. Kelly D. Jaundice in the neonate. Med Internat 1994; 22:461–8.
4. Roberts EA. The jaundiced baby. In: Kelly DA, ed. Diseases of the liver and biliary system in children. London: Blackwell Science, 1999: 11–46.
5. Davenport M, Howard E. Surgical disorders of the liver and bile ducts. In: Kelly DA, ed. Diseases of the liver and biliary system in children. London: Blackwell Science, 1999:253–78.

THIS CASE WAS PRESENTED BY DR. MICHELLE BIGNELL AND DR. DEIRDRE A. KELLY, BIRMINGHAM CHILDREN'S HOSPITAL, BIRMINGHAM, UK

22. Familial chronic fatigue

PRESENTATION

INITIAL PRESENTATION
- A 53-y-old woman presented to her primary care physician with a long history of profound lethargy associated with insomnia and arthralgia mainly affecting the knees
- The patient dated her symptoms to a flulike illness 6 mo previously

PREVIOUS MEDICAL HISTORY
- Hypertension treated with an angiotensin-converting enzyme inhibitor and thiazide diuretic
- She had also been taking estrogen replacement since menopause 2 y earlier
- She has been a blood donor until 13 y previously, donating a total of 24 units of blood
- She drank four units of alcohol per week, but did not smoke

PHYSICAL EXAMINATION
- Her physical examination produced no abnormal findings

INITIAL INVESTIGATIONS

Serum

Sodium	141 mEq/L	(141 mmol/L)
Potassium	3.8 mEq/L	(3.8 mmol/L)
Bicarbonate	24 mEq/L	(24 mmol/L)
Urea	8 mg/dL	(3.0 mmol/L)
Creatinine	1.0 mg/dL	(90 µmol/L)
Total thyroxine	7.7 µg/dL	(99 nmol/L)
Thyroid-stimulating hormone	3.0 µU/mL	(3.0 mU/L)
Cholesterol (random)	170 mg/dL	(4.4 mmol/L)
Calcium	9.7 mg/dL	(2.42 mmol/L)
Uric acid	5.1 mg/dL	(0.30 mmol/L)
Rheumatoid factor	<20 IU/L	(<20 IU/L)
Viral serology	Increased immunoglobulin G titer to Epstein-Barr virus	
Blood glucose (random)	90 mg/dL	(5.0 mmol/L)

X-rays of the knee joints were normal

- Two years after her initial presentation, her brother, who had also been suffering from longstanding fatigue, was diagnosed as having liver disease

---

WHAT ARE YOUR PROVISIONAL DIAGNOSES?

WHAT FURTHER INVESTIGATIONS WOULD YOU SUGGEST?

PROVISIONAL DIAGNOSES
- Chronic fatigue syndrome
- Liver disease?

FURTHER PROGRESS
- A diagnosis of chronic fatigue syndrome was made
- Over the following months her symptoms impaired her ability to work, shop, and perform household tasks
- Further medical consultations revealed no new features or abnormal tests, and she took early retirement on the grounds of poor health
- Fatigue is a common symptom of liver disease, which may be disabling and is often overlooked
- Although arthralgia is not a typical feature of chronic liver disease in general, fatigue and arthralgia are the most common presenting symptoms of genetic hemochromatosis, which is one of the most common autosomal recessive disorders
- Unlike the situation in almost every other type of acute or chronic liver disease, liver function tests are often normal in hemochromatosis
- Ultrasound scanning is usually not helpful in making the diagnosis, though hepatomegaly may be present
- Fasting transferrin saturation and serum ferritin should be performed if the diagnosis is suspected

FURTHER INVESTIGATIONS
- A fasting blood sample was collected and provided the following results:

Serum

Bilirubin	0.9 mg/dL	(16 µmol/L)
Alkaline phosphatase	118 IU/L	(118 IU/L)
Alanine aminotransferase	32 IU/L	(32 IU/L)
Albumin	4.0 g/dL	(40 g/L)
Total protein	7.2 g/dL	(72 g/L)
Iron	294 µg/dL	(52.6 µmol/L)
Total iron-binding capacity	350 µg/dL	(62.6 µmol/L)
Ferritin	2823 ng/mL	(2823 µg/L)
Transferrin saturation	84%	

WHAT IS YOUR DIAGNOSIS?

HOW WOULD YOU CONFIRM THIS DIAGNOSIS?

- A fasting transferrin saturation of 45% or more is typically used as a screening threshold for hemochromatosis, because it identifies 98% of affected persons while producing relatively very few false-positives
- When serum transferrin saturation is ≥55%, a serum ferritin of >200 μg/L in premenopausal women or >300 μg/L in men and postmenopausal women indicates primary iron overload due to hemochromatosis
- Serum ferritin can be normal early in the disorder before significant iron overload develops (1)
- If serum ferritin but not transferrin saturation is elevated, then hemochromatosis is unlikely
- Serum ferritin can be elevated in inflammatory and carcinomatous conditions, as well as liver diseases other than hemochromatosis, such as alcoholic liver disease

FINAL DIAGNOSIS: Genetic hemochromatosis (GH)

CONFIRMATION OF DIAGNOSIS

- The diagnosis is confirmed by gene testing and liver biopsy. More than 90% of cases of genetic hemochromatosis are the result of being homozygous for a C282Y mutation in the *HFE* gene on chromosome 6 (2)
- Most of the remaining cases are compound heterozygotes for this mutation and another mutation, H63D, in the same *HFE* gene. However lack of these mutations does not exclude the presence of significant iron overload, and liver biopsy may be indicated. Histology of sections stained with Perls' stain shows the characteristic pattern of parenchymal iron overload. Measurement of the dry liver iron weight, if available, is also valuable because the "hepatic iron index" can be calculated

$$\text{Hepatic iron index} = [\text{dry liver iron weight (μmol/g)}]/\text{age (y)}$$

- A value >2 reliably differentiates hemochromatosis from other conditions associated with increased hepatic iron stores (3)

MANAGEMENT AND SUBSEQUENT PROGRESS

- Treatment is with regular venesection, usually once per week, until iron stores are depleted. To achieve this, venesection should be continued until serum ferritin concentrations fall below 100 μg/L, and iron deficiency anemia develops
- Maintenance venesection is then required, usually two to four times per year. If cirrhosis is present, then there is a 200-fold increase in risk of the patient developing a hepatocellular carcinoma, even if iron stores are depleted (4)
- Serum α-fetoprotein concentrations are often not elevated in hepatocellular carcinoma associated with hemochromatosis, but regular ultrasound scanning in combination with α-fetoprotein testing, particularly in male cirrhotics, may detect tumors small enough for treatment
- Dietary restrictions are not generally necessary, but patients should be told to avoid vitamin preparations containing iron and vitamin C
- Screening of first-degree relatives with gene testing, fasting serum transferrin saturation, and serum ferritin should be undertaken. The heterozygote frequency of the abnormal gene is ~10% in the general population, and so sons and daughters of patients with hemochromatosis have a 5% chance of being homozygotes
- The patient had 65 units of blood (16 g of iron stores) removed, with a substantial improvement in her symptoms

KEY FEATURES
- Genetic hemochromatosis is an autosomal recessive inherited disorder of iron metabolism involving an inappropriately high level of iron absorption from the small bowel. It is the commonest inherited liver disorder and one of the commonest autosomal recessive disorders occurring in ~5 per 1000 Caucasian population (5)
- Untreated, iron gradually accumulates in a variety of tissues, and can eventually result in organ failure and the variable clinical signs and symptoms of the disease. The classical triad of pigmentation, diabetes mellitus, and cirrhosis, however, represents only the most severe end of a spectrum of clinical disease expression and hemochromatosis should be considered in the differential diagnosis of a wide variety of symptoms and presentation as shown in the table
- Asymptomatic cases are now often diagnosed by family screening

TABLE
Consider genetic hemochromatosis in the differential diagnosis of these conditions

• Liver disease	• Arthritis
• Heart failure	• Hypogonadism
• Hyperglycemia	• Unexplained fatigue

- Fatigue is the commonest symptom present at diagnosis regardless of whether cirrhosis is present or not. Although a symptom of liver failure and cirrhosis, fatigue is often a prominent syndrome of pre-cirrhotic hemochromatosis with normal liver function, suggesting that it is iron overload *per se* that causes this symptom. Liver function tests are often normal, with one study showing two-thirds of precirrhotic cases and one-third of cirrhotic cases having normal tests at diagnosis (4)
- Several studies have examined the usefulness of a variety of investigations in the assessment of patients presenting with chronic fatigue in primary care. In general such investigations have not proved useful as only a low yield of abnormal results has been found. However, such studies have not included screening tests for hemochromatosis and sometimes not even liver function tests
- Iron stores in hemochromatosis are promoted by a diet rich in red meat and vitamin C and depleted by any cause of blood loss. It is often wrongly thought that premenopausal women are protected from iron overload and organ damage by menstruation and the requirements of pregnancy. However this patient, who had cirrhosis due to hemochromatosis at diagnosis only four years after the menopause, illustrates that even combined blood loss due to blood donation and menstruation may not prevent iron accumulation and subsequent organ damage
- Venesection is required to deplete and maintain iron stores at the lower limit of normal. Although cirrhosis cannot be reversed and the risk of hepatocellular carcinoma cannot be removed once cirrhosis has developed, liver function, cardiac function and diabetic control may all improve with venesection as can symptoms of fatigue (4)

POINTS TO REMEMBER
- Genetic hemochromatosis is a relatively common disorder variable in its manifestations
- Pre- and perimenopausal women are not exempt from clinical expression of the disease
- Fasting serum transferrin saturation and serum ferritin are required to make a provisional diagnosis
- Testing for mutations in the HFE gene can confirm the diagnosis and allow accurate family studies
- Liver biopsy may be required to confirm the diagnosis and assess liver architecture

REFERENCES

1. Powell LW, George DK, McDonnell SM, Kowdley KV. Diagnosis of hemochromatosis. Ann Intern Med 1998;129:925–31.
2. Feder JN, Gnirke A, Thomas W, et al. A novel MHC class I like gene is mutated in patients with hereditary hemochromatosis. Nature Genet 1996;13:399–408.
3. Basset ML, Halliday JW, Powell LW. Value of hepatic iron measurements in early hemochromatosis and determination of the critical iron level associated with fibrosis. Hepatology 1986;6:24–9.
4. Niederau C, Fisher R, Sonnenberg A, et al. Survival and causes of death in cirrhotic and non-cirrhotic patients with primary hemochromatosis. N Engl J Med 1985;313:1256–62.
5. Phatak PD, Sham RL, Raubertas RF, et al. Prevalence of hereditary hemochromatosis in 16031 primary care patients. Ann Intern Med 1998;129:954–61.
6. Worwood M, What is the role of genetic testing in diagnosis of haemachromatosis. Ann Clin Biochem 2001;38:3–19.

THIS CASE WAS PRESENTED BY DR. IAN R. GUNN, LAW HOSPITAL, CARLUKE, LANARKSHIRE, UK

23. A misleading case of abdominal pain

PRESENTATION

HISTORY OF PRESENTING COMPLAINT
- A 28-y-old Asian woman presented with a 3-wk history of anorexia, generalized colicky abdominal pain, and constipation
- She had been able to eat and drink and had not vomited
- She had not had any rectal bleeding
- Her last menstrual period had been 10 d previously and had been normal

PAST MEDICAL HISTORY
- She was awaiting in vitro fertilization treatment for primary infertility

DRUG HISTORY
- She was not receiving any prescribed medication

FAMILY AND SOCIAL HISTORY
- She was a housewife and had been married for 10 y
- She did not smoke and did not drink alcohol

ON EXAMINATION
- She was apyrexial
- She appeared clinically anemic
- Blood pressure was 112/76 mmHg and pulse 88 beats/min and regular
- She had vague generalized abdominal tenderness, but no guarding or rebound tenderness
- Normal bowel sounds were present
- Cardiovascular, respiratory, and neurological examination revealed no other abnormalities

---

WHAT ARE YOUR PROVISIONAL DIAGNOSES?

WHAT INVESTIGATIONS ARE INDICATED IN THIS PATIENT?

PROVISIONAL DIAGNOSES
- Gastritis
- Gall bladder disease
- Peptic ulcer disease
- Appendicitis
- Pancreatitis
- Pelvic inflammatory disease

INITIAL INVESTIGATIONS

Sodium	138 mEq/L	(138 mmol/L)
Potassium	4.0 mEq/L	(4.0 mmol/L)
Urea	14 mg/dL	(5.0 mmol/L)
Creatinine	0.93 mg/dL	(82 µmol/L)
Glucose	105 mg/dL	(5.8 mmol/L)
Bilirubin	0.7 mg/dL	(12 µmol/L)
Albumin	3.8 g/dL	(38 g/L)
Aspartate aminotransferase	30 IU/L	(30 IU/L)
Alkaline phosphatase	76 IU/L	(76 IU/L)
Hemoglobin	8.0 g/dL	(80 g/L)
Leukocyte count	$5.5 \times 10^3/\mu L$	($5.5 \times 10^9/L$)
Platelets	$300 \times 10^3/\mu L$	($300 \times 10^9/L$)
Blood film	Normochromic normocytic	
	Basophilic stippling of erythrocytes	
Amylase	160 IU/L	
Hepatitis serology	Negative	
Pregnancy test	Negative	

- Abdominal X-ray showed fine radiodense stippling throughout the small and large bowel
- Abdominal ultrasound showed no abnormality
- Endoscopy showed normal esophagus, stomach, and duodenum
- The patient's husband then told a junior doctor that his wife's symptoms began soon after she started taking a traditional Asian medicine. This was a white powder that was supposed to be a treatment for infertility
- X-ray of the powder showed radiodense stippling corresponding to that seen on the abdominal X-ray

---

WHAT FURTHER INVESTIGATIONS, IF ANY, WOULD YOU REQUEST?

WHAT CHANGES HAVE YOU MADE TO YOUR PROVISIONAL DIAGNOSIS?

WHAT IS YOUR DIAGNOSIS?

A MISLEADING CASE OF ABDOMINAL PAIN

FURTHER INVESTIGATIONS

Blood lead	70 μg/dL	(3.5 μmol/L)
Erythrocyte zinc protoporphyrin	109 μg/dL	(1.9 μmol/L)

Analysis of the white powder showed it contained 12% lead by weight

TREATMENT
- She was given intravenous calcium edetate to chelate the lead. Her symptoms resolved within a week of starting treatment

FINAL DIAGNOSIS : Lead poisoning due to ingestion of a lead-containing traditional Asian remedy

KEY FEATURES

PATHOPHYSIOLOGY AND CLINICAL FEATURES
- Lead inhibits heme synthesis (see figure), causing anemia
- Porphyrins accumulate and cause peripheral or autonomic neuropathy. The latter can result in abdominal pain and constipation
- Encephalopathy is sometimes seen in severe poisoning
- Developmental delay may result from low-level exposure in children
- Renal tubular damage may also occur

FIGURE   THE MAJOR EFFECTS OF LEAD ON HEME SYNTHESIS

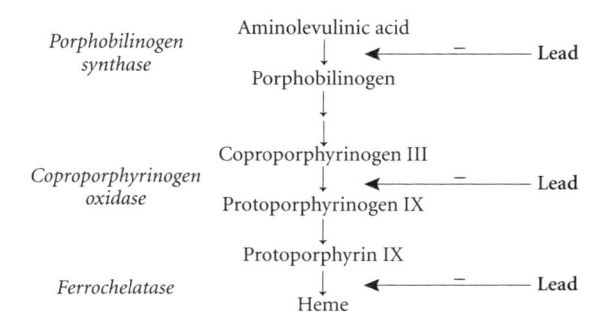

```
                          Aminolevulinic acid
      Porphobilinogen            ↓         ←  −  ──── Lead
         synthase          Porphobilinogen
                                  ↓
                                  ↓
                          Coproporphyrinogen III
    Coproporphyrinogen            ↓         ←  −  ──── Lead
         oxidase           Protoporphyrinogen IX
                                  ↓
                           Protoporphyrin IX
       Ferrochelatase             ↓         ←  −  ──── Lead
                                Heme
```

LABORATORY FINDINGS
- Blood film shows "basophilic stippling" of erythrocytes due to inhibition of pyrimidine 5′ nucleotidase and accumulation of aggregates of RNA
- Erythrocyte zinc protoporphyrin accumulates when inhibition of ferrochelatase (see figure) prevents iron incorporation into porphyrin. There is nonenzymatic incorporation of zinc
- Blood lead concentration has fallen with the use of unleaded gasoline and should be <10 μg/dL (<0.5 μmol/L). Concentrations between 10 and 20 μg/dL (0.5 and 1.0 μmol/L) may be associated with a slight reduction in I.Q. in children, but features of acute poisoning in adults are rarely seen unless blood lead concentration exceeds 60 μg/dL (3.0 μmol/L)

ETIOLOGY
- Common causes are occupational exposure in adults and ingestion of lead-containing paint in children with pica
- Less common are poisoning from exposure to lead solder or domestic water pipes, ceramic glazes, and traditional Chinese and other Asian cosmetics and remedies

POINTS TO REMEMBER
- Consider lead poisoning as a rare cause of abdominal pain, peripheral neuropathy, elevated porphyrins, or anemia
- Blood lead concentration is the preferred investigation for suspected inorganic lead poisoning, and urinary lead excretion for cases of organic lead exposure
- Zinc protoporphyrin is a useful marker of lead exposure, but is also elevated in iron-deficiency anemia
- Chinese and South East Asian traditional remedies contain heavy metals as intentional constituents. The commonest are lead, mercury, and arsenic. Doses prescribed can cause symptomatic poisoning

REFERENCES
1. Ko RJ. Adulterants in Asian patent medicines. New Engl J Med 1998;339:847.
2. Bayly GR, Braithwaite RA, Sheehan TMT, Dyer NH, Grimley C, Ferner RE. Lead poisoning from Asian traditional remedies in the West Midlands—report of a series of five cases. Hum Exp Toxicol 1995;14:24–8.
3. Baldwin DR, Marshall WJ. Heavy metal poisoning and its laboratory investigation. Ann Clin Biochem 1999;36:267–300.

THIS CASE WAS PRESENTED BY DR. GRAHAM BAYLY, BRISTOL ROYAL INFIRMARY, BRISTOL, UK

24. Neonatal hypocalcemia

PRESENTATION

INITIAL PRESENTATION
- At birth, the patient was a boy weighing 7.5 pounds (3.4 kg), condition poor, heart rate <100 beats/min, cyanosed, no respiratory effort
- Unresponsive, floppy, dysmorphic features noted
- Oral pharyngeal suction, bag and mask ventilation, intubation unsuccessful
- At 10 min: Pink, good heart rate, regular respiration, still floppy
- Transferred to neonatal unit and intubated
- Neurology: Jerky movements, irritable
- Moderate to severe hypoxic ischemic encephalopathy

MOTHER'S PAST MEDICAL HISTORY
- Mother was a 28-y-old Bangladeshi
- Gravida 7; five healthy children, all female, one miscarriage
- This pregnancy, she developed gestational diabetes (requiring insulin) and polyhydramnios
- Labor induced at 37 wk by artificial rupture of membranes
- Delivery precipitate

ON EXAMINATION
- Abnormal low-set ears
- Flat bridge of nose
- Large cleft palate
- Micropenis (testes in scrotum)
- Cardiovascular system (day 2): Sharp ejection systolic murmur detected

INITIAL INVESTIGATIONS

Day 1

Serum calcium	7.6 mg/dL	(1.90 mmol/L)
Corrected calcium	8.2 mg/dL	(2.06 mmol/L)
Phosphate	3.6 mg/dL	(1.17 mmol/L)
Alkaline phosphatase	111 IU/L	(111 IU/L)

Day 2

Serum calcium at 0200 h	7.4 mg/dL	(1.84 mmol/L)
Serum calcium at 1400 h	6.2 mg/dL	(1.56 mmol/L)

The baby was treated with intravenous (i.v.) calcium and magnesium

---

WHAT ARE YOUR PROVISIONAL DIAGNOSES?

WHAT FURTHER INVESTIGATIONS WOULD YOU REQUEST?

PROVISIONAL DIAGNOSES
- Physiological hypocalcemia
- Hypomagnesemia
- Vitamin D deficiency
- Congenital hypoparathyroidism
- Pseudohypoparathyroidism

FURTHER INVESTIGATIONS
- Computed tomography scan:
  - Abnormal ductus between aorta and pulmonary artery (interrupted aortic arch)
  - Patent foramen ovale (R-L shunting)
  - Patent ductus arteriosis (bidirectional shunting)
  - Absent thymus
- Baby on total parenteral nutrition (TPN) receiving i.v. calcium days 2–12

PROGRESS

DAY	CALCIUM [mg/dL (mmol/L)]	CORRECTED CALCIUM [mg/dL (mmol/L)]	PHOSPHATE [mg/dL (mmol/L)]
3	7.1 (1.76)	7.5 (1.88)	5.9 (1.92)
4	5.7 (1.43)	6.3 (1.57)	–
5	7.1 (1.78)	–	–
6	10.2 (2.54)	–	–
7	10.0 (2.50)	10.5 (2.62)	–
8	6.3 (1.58)	7.1 (1.78)	4.6 (1.48)
9	8.9 (2.23)	9.8 (2.45)	–
10	8.6 (2.16)	9.5 (2.38)	–
11	8.5 (2.12)	–	–

- Magnesium was not measured
- Baby weaned from TPN; i.v. calcium changed to oral calcium

DAY	CALCIUM [mg/dL (mmol/L)]	CORRECTED CALCIUM [mg/dL (mmol/L)]	PHOSPHATE [mg/dL (mmol/L)]	PTH [pg/mL (pmol/L)]	25-OH-D [ng/mL (nmol/L)]
13	7.9 (1.98)	–	–	–	–
14	8.6 (2.16)	9.4 (2.34)	–	–	–
15	7.9 (1.98)	8.7 (2.18)	–	–	–
16	7.0 (1.74)	7.8 (1.94)	–	–	–
17	5.2 (1.31)	5.9 (1.47)	7.2 (2.32)	0.7 (0.74)	104 (40)
18	5.0 (1.24)	5.6 (1.40)	–	–	–
19	5.2 (1.29)	5.7 (1.43)	–	–	–
21	5.2 (1.31)	5.8 (1.45)	–	–	–

PTH, parathyroid hormone; 25–OH–D, 25–hydroxyvitamin D

WHAT IS YOUR FINAL DIAGNOSIS?

WHAT ARE THE CAUSES OF NEONATAL HYPOCALCEMIA?

WHAT IS THE NORMAL COURSE OF SERUM CALCUM IN THE NEONATAL PERIOD?

NEONATAL HYPOCALCEMIA 115

- The persistence of the hypocalcemia suggests that this is not physiological
- The requirement for i.v. calcium rather than oral calcium suggests a problem with calcium absorption
- There are adequate concentrations of 25-hydroxyvitamin D; hence, this is not due to vitamin D deficiency
- The concentration of parathyroid hormone (PTH) is inappropriate for the low calcium, confirming hypoparathyroidism rather than pseudohypoparathyroidism
- In this case, the hypoparathyroidism coupled with absent thymus led to a diagnosis of DiGeorge syndrome
- The other dysmorphic features present in this baby led to a diagnosis of CHARGE association (see Key Features) with DiGeorge syndrome

FINAL DIAGNOSIS: DiGeorge syndrome

FURTHER PROGRESS
- Day 42: Treated with α_1-hydroxyvitamin D
- Day 43: Calcium 5.88 mg/dL (1.47 mmol/L)
- Day 46: Calcium 8.24 mg/dL (2.06 mmol/L); corrected calcium 8.8 mg/dL (2.20 mmol/L)
- Baby maintained on α_1-hydroxyvitamin D with normal calcium concentrations since then

POSSIBLE CAUSES OF NEONATAL HYPOCALCEMIA
- Physiological (early onset)
  - Prematurity
  - Maternal diabetes
  - Birth asphyxia
- Pathological
  - Maternal vitamin D deficiency or malabsorption
  - Hypomagnesemia
  - Liver or renal disease
  - Congenital hypoparathyroidism
  - Pseudohypoparathyroidism
- Iatrogenic
  - Low calcium intake
  - High phosphate intake
  - Anticonvulsants

WHAT IS THE NORMAL COURSE OF SERUM CALCIUM IN THE EARLY NEONATAL PERIOD?
- Term infants are born with high serum calcium concentrations due to massive accretion of calcium by the fetus in the third trimester of pregnancy. Premature infants have lower calcium values because the period for calcium accretion is reduced—the more premature the infant, the lower the calcium
- Fetal accretion of calcium is due to active transplacental transport of calcium by 1,25-dihydroxyvitamin D synthesized by the placenta. Hence, at birth the infant's own PTH is suppressed, and 1,25-dihydroxyvitamin D is low. Serum calcium concentrations fall rapidly, reaching a nadir at day 2
- Some degree of hypocalcemia is normal and may be exacerbated by prematurity or by stress release of phosphate from tissues (maternal diabetes, birth asphyxia), but is usually self-limiting

because falling calcium concentrations promote PTH secretion and 1,25-dihydroxyvitamin D production

KEY FEATURES
- DiGeorge syndrome described in 1965
  - Hypoparathyroidism
  - Cellular immune deficiency
  - Absent thymus and parathyroids
- Subsequently, it has been shown to be a more variable syndrome, with many patients having:
  - Cardiovascular anomalies
  - Dysmorphic facies (such as low-set pixie ears)
  - Other anomalies not related to the anterior neck and thorax
- The most common presenting feature was congenital heart disease; there was a high association with interrupted aortic arch
- Revised criteria: Two of the following = DiGeorge syndrome
  - Cellular immune deficiency or absent thymus or both
  - Symptomatic hypocalcemia or anatomical deficiency of parathyroids or both
  - Congenital heart disease. Modified to "if there is no thymic or parathyroid anomaly the cardiac defect must be aortic arch anomaly"
- Genetics: Most cases of DiGeorge syndrome are sporadic, but there is a specific association with deletion of the proximal portion of chromosome 22 in 10% of cases. In this case, fluorescent in situ hybridization for DiGeorge–velocardiofacial syndrome deletion region of chromosome 22 was negative

CHARGE ASSOCIATION
- Coloboma: Cleft or failure to close eyeball, causing abnormalities of retina or optic nerve (present in this baby)
- Heart defects (present in this baby)
- Atresia of the choanae: The passages from the back of the nose to the throat are blocked (not present in this baby)
- Retardation of growth and development (difficult to assess neonatally)
- Genital defects (present in this baby)
- Ears: Unusual external ears (present in this baby)

- A significant number of patients with CHARGE association also have DiGeorge syndrome

POINTS TO REMEMBER
- Total calcium concentrations require measurement of albumin for interpretation (corrected calcium). This indirect attempt to assess the ionized fraction may not always be reliable in babies
- In term infants, serum calcium is higher than normal adult concentrations at birth, but falls rapidly (days 1–2), then rises to the adult concentration (by day 3)
- Physiological neonatal hypocalcemia is exacerbated by prematurity, birth asphyxia, and maternal diabetes and may be symptomatic
- Persistent neonatal hypocalcemia is uncommon and requires further investigation
- Neonatal hypocalcemia in combination with dysmorphic features and other abnormalities (particularly cardiac) may indicate a congenital syndrome or association such as DiGeorge and/or CHARGE

REFERENCES

1. Tietz NW, ed. Clinical guide to laboratory tests, 2nd ed. Philadelphia: WB Saunders, 1990.
2. Mayne PD, Kovar IZ. Calcium and phosphorus metabolism in the premature infant. Ann Clin Biochem 1991;28:131–41.
3. Conley ME, Beckwith JB, Mancer JFK, Tenckhoff L. The spectrum of the DiGeorge syndrome. J Pediatr 1979;94:883–80.
4. Carey JC. Spectrum of the DiGeorge "syndrome" [Editorial correspondence]. J Pediatr 1980;96:955–6.
5. Pagon RA, Graham JM, Zonana J, Yong SL. Coloboma, congenital heart disease, and choanal atresia with multiple anomalies: CHARGE association. J Pediatr 1981;99:223–7.

THIS CASE WAS PRESENTED BY DR. CATHERINE A. STREET, ROYAL LONDON HOSPITAL, LONDON, UK

25. Unusual biochemical changes after a flu-like illness

PRESENTATION

HISTORY
- A 32-y-old man presented with a 4-d history of an influenza-like illness accompanied by polyuria and polydipsia
- History was elicited from his wife because the patient was confused and aggressive

PAST MEDICAL HISTORY
- Nothing significant
- No history of diabetes

FAMILY HISTORY
- No family history of diabetes or any other disease

SOCIAL HISTORY
- Occasional social drinker
- A nonsmoker with no history of drug abuse
- He was sedated with chlorpromazine, haloperidol, and diazepam

ON EXAMINATION
- Semiconscious and rolling about on hospital trolley; confused
- No focal neurological signs
- No significant abnormalities found
- Breathing normally
- Dehydrated but well perfused

WHAT ARE YOUR PROVISIONAL DIAGNOSES?

WHAT INVESTIGATIONS ARE INDICATED IN THIS PATIENT?

PROVISIONAL DIAGNOSES
- Metabolic problem such as diabetic ketoacidosis
- Poisoning
- Viral encephalopathy

INITIAL INVESTIGATIONS
Plasma

Sodium	145 mEq/L	(145 mmol/L)
Potassium	4.5 mEq/L	(4.5 mmol/L)
Bicarbonate	22 mEq/L	(22 mmol/L)
Urea	87 mg/dL	(31 mmol/L)
Glucose	1485 mg/dL	(82.4 mmol/L)
Osmolality	413 mOsm/kg	(413 mOsm/kg)

The blood sample was reported as hemolyzed
Urine analysis: 2% glucose; ketones positive; blood negative

WORKING DIAGNOSIS
- Hyperosmolar diabetic decompensation

MANAGEMENT
- Treated with intravenous isotonic saline, intravenous insulin, potassium, and subcutaneous heparin
- Glucose fell to 234 mg/dL (13 mmol/L) over the next 36 h while his sodium rose to 182 mEq/L (182 mmol/L). His plasma osmolality over this period did not alter greatly but declined slowly over the next 2 d
- He developed septicemia and a tense abdomen, possibly due to pancreatitis (amylase 3480 IU/L)
- Laboratory contacted on day 6 because of a persistent acidosis [bicarbonate 16 mEq/L (16 mmol/L)] despite being well perfused. The acidosis was of the hyperchloremic variety, and the patient had a normal anion gap
- Urine, which had been red for some days, was positive for myoglobin

WHAT IS YOUR FINAL DIAGNOSIS?

FINAL DIAGNOSIS: Rhabdomyolysis complicating hyperosmolar diabetic decompensation

FURTHER PROGRESS
- Plasma creatine kinase (day 6) was 332,000 IU/L, and a diagnosis of rhabdomyolysis was confirmed
- His plasma calcium at this time was 6.3 mg/dL (1.58 mmol/L) (corrected for albumin)
- His renal function declined and he suffered a fatal cardiorespiratory arrest. No viral cause for the rhabdomyolysis was found, and autopsy showed necrosis of skeletal and cardiac muscle and of the renal tubules

PATHOPHYSIOLOGY
- Plasma osmolality declined slowly since the falling glucose was accompanied by a rise in sodium:
  - Seen frequently in hyperosmolar coma because the raised glucose draws water out of cells
  - On treatment with insulin, glucose enters cells accompanied by water and reveals hypernatremia. A correction for this has been suggested; if no water movement had occurred:

  Corrected sodium = glucose/4 + measured sodium (in mmol/L)

- The raised amylase may have reflected pancreatitis, but it can be elevated by nonpancreatic causes such as diabetic ketoacidosis, frequently to 5–10 times the upper reference range
- The persistent metabolic acidosis accompanied the rhabdomyolysis in which myoglobin was deposited in the renal tubules
  - Predominantly tubular damage (as opposed to glomerular) is associated with a metabolic acidosis due to reduced aldosterone responsiveness of the renal tubule in which acid anions are excreted normally but hydrogen ions are retained
- Sixty percent of cases of rhabdomyolysis have associated hypocalcemia, which may be due to sequestration of plasma calcium by damaged muscle

POINTS TO REMEMBER
- Although unusual, hyperosmolar decompensation can present in younger age groups
- Hypernatremia may develop during the treatment of hyperosmolar diabetic decompensation
- Increased plasma amylase activity may occur in nonpancreatic disease
- The calculation of the anion gap may be helpful in the elucidation of metabolic acidosis of unknown origin
- In renal failure, predominantly tubular damage may lead to normal anion gap acidosis
- Rhabdomyolysis is frequently accompanied by hypocalcemia

REFERENCES
1. Walmsley RN, White GH. A Guide to diagnostic clinical chemistry, 3rd ed. Oxford, UK: Blackwell Scientific Publications, 1994.
2. Gabow PA, Kaehny WD, Kelleher SP. The spectrum of rhabdomyolysis. Medicine 1982;61:141–52.
3. Llach F, Felsenfeld AJ, Haussler MR. The pathophysiology of altered calcium metabolism in rhabdomyolysis-induced acute renal failure. N Engl J Med 1981;305:117–23.
4. Hooper J, Wood MLB. Some unusual biochemical changes in diabetes mellitus. J R Soc Med 1996;89:218P–20P.

THIS CASE WAS PRESENTED BY DR. JAMES HOOPER, ROYAL BROMPTON HOSPITAL, LONDON, UK, AND IS BASED ON HOOPER J, WOOD MLB. SOME UNUSUAL BIOCHEMICAL CHANGES IN DIABETES MELLITUS. J R SOC MED 1996;89:218P–20P.

26. A suspected overdose

PRESENTATION

HISTORY OF PRESENTING COMPLAINT
- A 29-y-old woman was admitted to the emergency department, accompanied by a friend, at 2345 h with a recent history of heavy drinking over the course of the past few hours
- The friend had found the patient vomiting in the apartment they shared with two other people on her return from work earlier in the evening
- The friend had noticed a couple of tablets on the floor by the patient, but did not notice a bottle nearby

DIRECT QUESTIONING
- Although coherent, the patient was extremely distressed and cried at every attempt to ask questions
- Eventually the patient admitted to having been drinking vodka all day and having taken a quantity of "tablets"

PAST MEDICAL HISTORY
- Nothing of note

SOCIAL HISTORY
- The friend suggested recent difficulties in her relationship with partner of 3 y and problems at work

FAMILY HISTORY
- Nothing of note

ON EXAMINATION
- Her breath smelled of alcohol
- Respiration rate, pulse, and temperature were normal
- She had obviously been vomiting, but appeared to be reasonably well hydrated
- No other abnormalities were found

WHAT IS YOUR PROVISIONAL DIAGNOSIS?

WHAT INVESTIGATIONS WOULD YOU REQUEST?

PROVISIONAL DIAGNOSES
- Emotional crisis
- Excessive alcohol intake
- Drug overdose

INITIAL INVESTIGATIONS
Serum

Sodium	142 mEq/L	(142 mmol/L)
Potassium	3.6 mEq/L	(3.6 mmol/L)
Bicarbonate	20 mEq/L	(20 mmol/L)
Urea	15 mg/dL	(5.4 mmol/L)
Creatinine	1.0 mg/dL	(90 µmol/L)
Acetaminophen (paracetamol)	8.3 mg/dL	(0.55 mmol/L)
Salicylate	None detected	

WHAT FURTHER INVESTIGATIONS, IF ANY, WOULD YOU REQUEST?

HAVE YOU MADE ANY CHANGES TO YOUR PROVISIONAL DIAGNOSIS?

WHAT IS YOUR DIAGNOSIS?

WORKING DIAGNOSIS
- Excessive intake of acetaminophen and alcohol

FURTHER INVESTIGATIONS
- Biochemical tests to assess liver dysfunction immediately, with follow-up in 2 and 7 d time
- Prothrombin time
- Initial results were as follows:

    Serum

Bilirubin	0.7 mg/dL	(12 µmol/L)
Alanine aminotransferase	40 IU/L	(40 IU/L)
Alkaline phosphatase	90 IU/L	(90 IU/L)
Albumin	4.1 g/dL	(41 g/L)
Prothrombin time	18 s	

- Two days later the results were as follows:

    Serum

Sodium	144 mEq/L	(144 mmol/L)
Potassium	4.1 mEq/L	(4.1 mmol/L)
Bicarbonate	22 mEq/L	(22 mmol/L)
Urea	17 mg/dL	(6.1 mmol/L)
Creatinine	1.1 mg/dL	(97 µmol/L)
Bilirubin	1.5 mg/dL	(25 µmol/L)
Alanine aminotransferase	1972 IU/L	(1972 IU/L)
Alkaline phosphatase	97 IU/L	(97 IU/L)
Albumin	4.0 g/dL	(40 g/L)
Prothrombin time	22 s	

- Seven days later the alanine aminotransferase concentration had fallen to 62 IU/L and the bilirubin to 0.6 mg/dL (10 µmol/L), with the prothrombin falling to 15 s and the other results remaining unchanged

FINAL DIAGNOSIS: Acetaminophen poisoning with excessive alcohol intake

MANAGEMENT
- She was treated with an intravenous infusion of the antidote *N*-acetylcysteine. The dose was initially given at a high concentration, which was later reduced, the total period of the infusion being 15 h
- Some abdominal tenderness occurred over the next 2 d and then resolved
- She was encouraged to take plenty of liquids and to take food over the first 2 d, and her general state improved. She was encouraged to seek counseling

KEY FEATURES

PRESENTATION
- Initial presentation is often unremarkable with anorexia, nausea, and vomiting over the first 24 h
- Over the next 24 h abdominal pain and tenderness of the liver will develop
- Derangement of tests of liver function will develop within 48 h of ingestion of excessive amounts of acetaminophen with elevation of alanine aminotransferase and possibly bilirubin and a rise in the prothrombin time
- Careful questioning of the patient (and accompanying persons) will help to provide clues to a potential overdose

BIOCHEMISTRY
- Acetaminophen is metabolized by the liver, forming harmless conjugates (primarily glucuronides and sulfur-containing compounds), which are excreted by the kidney
- A small proportion of drug is metabolized to a highly toxic intermediate (*N*-acetyl-*p*-benzoquinoneimine, NAPQI), which is conjugated with glutathione to produce a nontoxic conjugate
- At higher levels of drug ingestion the glucuronidation and sulfation pathways become saturated, more NAPQI is produced, and the hepatic glutathione stores become depleted
- The toxic metabolite will bind strongly to hepatocellular proteins leading to cellular damage and interruption of metabolic processes and progressing rapidly to liver failure
- The toxic metabolite is also nephrotoxic, and exposure can lead to the development of renal failure

DIAGNOSIS
- The key action involves the recognition and quantification of the circulating drug concentration
- The requirement for treatment and the clinical outcome are related to the amount of parent drug present. Guidance is based on the nomogram illustrated in the figure
- Additional tests should include biochemical tests of liver dysfunction (particularly alanine aminotransferase), plasma creatinine, and prothrombin time to provide a baseline
- Repeating the above tests will indicate the progress of hepatic and renal function

TREATMENT
- The antidote most commonly used is *N*-acetylcysteine, which provides repletion of the hepatic glutathione stores, enabling the detoxification of NAPQI to a glutathionine conjugate
- Methionine has been used as an alternative antidote
- The patient should also be kept hydrated with the use of 5% dextrose
- Ingestion of excessive amounts of alcohol with the drug or the coexistence of alcoholic liver disease increases the risk of liver damage
- Treatment is most effective within the first few hours of ingestion and is much less effective if initiated >16 h after ingestion
- There have been cases of anaphylactic shock reported after administration of the antidote

POINTS TO REMEMBER
- Metabolism of acetaminophen taken in excessive amounts can lead to production of larger amounts of toxic metabolites that deplete the glutathione stores, resulting in hepatocellular damage and liver failure
- The antidote *N*-acetylcysteine restores the glutathione stores, enabling conjugation of the toxic metabolite and limiting the degree of hepatocellular damage
- Measurement of alanine aminotransferase and prothrombin time is advised to assess liver function together with plasma creatinine to assess renal function

FIGURE   NOMOGRAM RELATING PLASMA ACETAMINOPHEN CONCENTRATION TO OUTCOME, USED TO GUIDE NEED FOR THERAPY

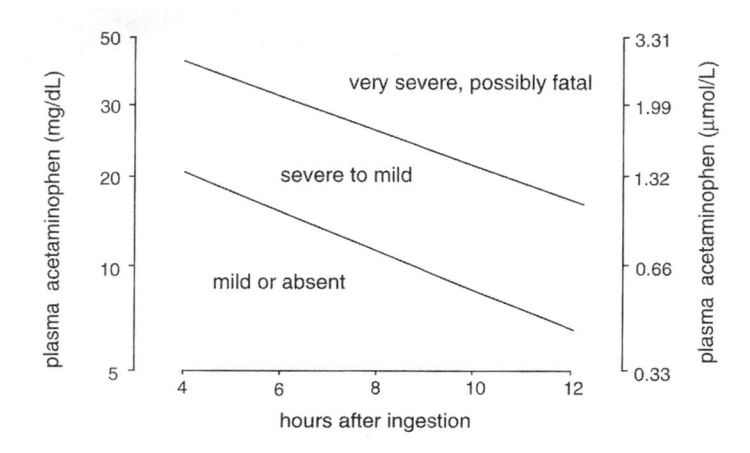

Adapted from Meredith TJ, Vale JA. Paracetamol poisoning. In: Vale JA, Meredith TJ, eds. Poisoning: diagnosis and treatment. London: Update Books, 1981:104–12.

REFERENCES

1. Bateman DN, Woodhouse KW, Rawlins MD. Adverse reactions to N-acetylcysteine. Lancet 1984;2:228.
2. Dresibach RH, Robertson WO, eds. Handbook of poisoning, 12th ed. Norwalk, CT: Appleton & Lange, 1987.
3. Marshall WJ, ed.. Clinical chemistry, 4th ed. London: Mosby, 2000.
4. Meredith TJ, Vale JA. Paracetamol poisoning. In: Vale JA, Meredith TJ, eds. Poisoning: diagnosis and treatment. London: Update Books, 1981:104–12.
5. Prescott LF, Critchley JA. The treatment of acetaminophen poisoning. Annu Rev Pharmacol Toxicol 1983;23:87–101.
6. Prescott LF, Illingworth RN, Critchley JAJH, et al. Intravenous N-acetylcysteine: the treatment of choice for paracetamol poisoning. Br Med J 1979;2:1097–100.
7. Renzi FP, Donovan JW, Martin TG, Morgan L, Harrison EF. Concomitant use of activated charcoal and N-acetylcysteine. Ann Emerg Med 1985;14:568–72.

THIS CASE WAS PRESENTED BY PROFESSOR CHRISTOPHER P. PRICE, THE ROYAL LONDON HOSPITAL, BARTS AND LONDON NHS TRUST, LONDON, UK

27. Prenatal screening and MoMs

PRESENTATION

- Patient presented for routine second-trimester maternal serum screening
- She was 16 wk 4 d gestation by dating scan
- She weighed 194 pounds (88 kg) and had no previous history of a Down's syndrome or a neural-tube defect pregnancy
- She did not have type 1 diabetes mellitus
- Her blood was analyzed for α-fetoprotein (AFP), unconjugated estriol (uE_3), and human choriogonadotrophin (hCG), and the individual biochemical answers were converted into multiples of the median (MoM) based on her scan gestation

Her biochemical results were:

ANALYTE	RESULT	MoM
AFP	207 IU/mL	7.83
uE_3	2.0 ng/mL (7.1 nmol/L)	2.12
hCG	108.9 IU/mL	5.80

WHAT COULD HAVE CAUSED SUCH ABNORMAL RESULTS?

WHAT FURTHER INVESTIGATIONS ARE INDICATED IN THIS PATIENT?

POSSIBLE CAUSES OF ABNORMAL RESULTS
- Abnormal renal function
- Abnormal liver function
- Interferences with the assay
- Multiple pregnancy
- Acute parvovirus B19 infection

FURTHER INVESTIGATIONS
Renal and liver function tests were within reference limits
Serum

Sodium	141 mEq/L	(141 mmol/L)
Potassium	3.6 mEq/L	(3.6 mmol/L)
Urea	5 mg/dL	(1.7 mmol/L)
Creatinine	0.7 mg/dL	(62 µmol/L)
Bilirubin	0.3 mg/dL	(5 µmol/L)
Albumin	3.3 g/dL	(33 g/L)
Total protein	6.0 g/dL	(60 g/L)
Alanine aminotransferase	18 IU/L	(18 IU/L)
Alkaline phosphatase	135 IU/L	(135 IU/L)
γ-Glutamyltransferase	17 IU/L	(17 IU/L)

- Assay of hCG and AFP on another system indicated a similar pattern of results with grossly elevated AFP and hCG results for a patient of 16–17 wk gestation
- Ultrasound scanning demonstrated a single fetus and no obvious abnormalities were detected
- Further blood tests were collected 3 d after the first test results were available
- Parvovirus B19 results were still pending at this stage. Komischke et al. (*2*) studied 65 patients with acute parvovirus B19 infection with fetal complications and found that 37% had elevated AFP results and 71% had elevated hCG results
- After the ultrasound scan, she was determined to be 18 wk gestation

Her new results, calculated at 18 wk gestation, were:

ANALYTE	RESULT	MoM
AFP	257 IU/mL	8.12
uE$_3$	2.5 ng/mL (8.7 nmol/L)	1.77
hCG	108.5 IU/L	7.13

WHAT IS YOUR DIAGNOSIS AT THIS STAGE?

RESULTS AND INTERPRETATION

- The pattern of results is similar to the first sample, so the patient was referred for an urgent detailed ultrasound scan. This was a difficult specialist examination, and eventually an acardiac twin was identified. Twin one appeared normal, and twin two was acardiac and acephalic with exomphalos
- An acardiac twin is a sporadic malformation confined to one of a set of monozygotic twins with a frequency of about 1 in 30,000 deliveries or 1 in 100 monozygotic twins. Mortality is 100% in the acardiac twin, and although the healthy twin is anatomically normal, the mortality for this twin is ~50%
- This mother was referred to a specialist hospital for laser treatment to remove the acardiac twin. After successful ablation by thermocoagulation of twin two's placental vessels, the pregnancy went well, and the mother delivered a healthy baby girl at 39 wk

FINAL DIAGNOSIS: Pregnant woman carrying twins, one of which was acardiac

BACKGROUND

- Second-trimester screening is usually carried out between 15 and 22 wk gestation
- Values for AFP and uE_3 rise with gestational age, whereas hCG values fall with gestation after 15 wk
- MoMs are used to express an individual patient's biochemical answer relative to a median value of 1.0 MoM; these are then used in conjunction with the patient's age, weight, and various other factors to calculate the Down's risk
- Median values are calculated from the biochemical results obtained from women with unaffected pregnancies
- In patients at high risk of having a Down's syndrome baby, the pattern of metabolites is usually a raised hCG with low AFP and uE_3
- A raised AFP result, usually >2.3 MoM, requires further investigation and is suggestive of a neural tube defect
- Second-trimester Down's screening may also detect babies at high risk of a range of other defects; for example, trisomy 18 pregnancies frequently have low results for all three metabolites

POINTS TO REMEMBER

- Appreciate that communication between laboratory and prenatal diagnosis staff is essential when running a maternal serum screening scheme
- Need to develop confidence in the laboratory results produced
- Know what abnormal results could represent
- Appreciate that without swift action and the removal of the acardiac twin, the healthy twin would almost certainly have died

REFERENCES

1. Wald NH, Kennard A, Hackshaw A, McGuire A. Antenatal screening for Down's syndrome. J Med Screen 1997;4:181–246.
2. Komischke K, Searle K, Enders G. Maternal serum alpha-fetoprotein and human chorionic gonadotropin in pregnant women with acute parvovirus B19 infection with and without fetal complications. Prenat Diagn 1997;17:1039–66.

THIS CASE WAS PRESENTED BY DR. GWYN MCCREANOR, KETTERING GENERAL HOSPITAL, KETTERING, UK

28. A difficult neurological case

PRESENTATION

HISTORY OF PRESENTING COMPLAINT
- A 38-y-old male gardener was brought to the emergency department after having collapsed unconscious at home. Immediately before this, he had had diplopia, an unsteady gait, and slurred speech
- His primary-care physician had called and found him to be unresponsive with tachycardia and unrecordably low blood pressure. He gave penicillin intravenously and called an ambulance

DIRECT QUESTIONING (HISTORY FROM WIFE)
- Diarrhea and vomiting for 1 wk
- Shingles for 3 d

PAST MEDICAL HISTORY
- Deaf since age 18
- Several admissions in the last 12 mo with odd behavior and one witnessed fit
- Computed tomography (CT) and magnetic resonance imaging scans had shown evidence of a left temporoparietal infarct from which he had made a full recovery

DRUGS
- Carbamazepine 600 mg, aspirin 150 mg

SOCIAL HISTORY
- Nonsmoker, occasional alcohol only

FAMILY HISTORY
- None of note

ON EXAMINATION
- On arrival, he was now conscious but confused, with a Glasgow Coma Score of 13/15
- Temperature 38 °C, blood pressure 95/50 mmHg, pulse 110/min
- No neck stiffness
- He had shingles in the T12 dermatome
- Neurological examination showed increased tone and brisk reflexes with an upgoing plantar response, all on the left side
- Funduscopy was not possible because of lack of cooperation

WHAT ARE YOUR PROVISIONAL DIAGNOSES?

WHAT INVESTIGATIONS ARE INDICATED?

PROVISIONAL DIAGNOSES
- Fit due to vomiting of carbamazepine
- New stroke
- Herpes zoster encephalitis
- Bacterial or viral meningitis

INITIAL INVESTIGATIONS

CT brain scan	No new abnormalities	
Lumbar puncture	Opening pressure 15 cm H_2O	
	Clear fluid. No leukocytes, no organisms	
Cerebrospinal fluid		
Protein	20 mg/dL	(200 mg/L)
Glucose	128 mg/dL	(7.1 mmol/L)
Serum		
Sodium	127 mEq/L	(127 mmol/L)
Potassium	4.6 mEq/L	(4.6 mmol/L)
Urea	34 mg/dL	(12.0 mmol/L)
Creatinine	2.4 mg/dL	(212 µmol/L)
Random plasma glucose	175 mg/dL	(9.7 mmol/L)
C-reactive protein	35.2 mg/dL	(352 mg/L))
Serum		
Carbamazepine	9.1 µg/mL	(38.5 µmol/L)

- The patient was given cefotaxime and acyclovir empirically and resuscitated with intravenous fluids. He was also started on subcutaneous heparin. He remained pyrexial and hypotensive. Ten days after admission he developed a grossly distended abdomen and peritonism with X-ray evidence of cecal dilation. He underwent an urgent colectomy and ileostomy. Histological examination showed ischemic colitis.
- Ten days postoperatively, his routine electrolytes showed:

Serum		
Sodium	118 mEq/L	(118 mmol/L)
Potassium	8.7 mEq/L	(8.7 mmol/L)
Bicarbonate	22 mEq/L	(22 mmol/L)
Urea	22 mg/dL	(7.9 mmol/L)
Creatinine	0.9 mg/dL	(81 µmol/L)

WHAT FURTHER INVESTIGATIONS, IF ANY, ARE REQUIRED?

WHAT IS THE IMMEDIATE MANAGEMENT?

- Electrocardiogram: Tented T waves with normal PR interval and QRS
- Repeat potassium confirmed the results (not taken from intravenous arm)
- Random plasma cortisol 26.3 µg/dL (726 nmol/L)
- Blood was taken for plasma renin activity and aldosterone assays
- He was given dextrose and insulin plus calcium gluconate intravenously stat
- Oral calcium resonium was started
- In addition, the heparin was stopped and fludrocortisone started

> WHAT WAS THE CAUSE OF THE HYPERKALEMIA?

WORKING DIAGNOSIS
- Hypoaldosteronism secondary to prolonged heparin therapy

PROGRESS
- His serum potassium fell rapidly over the next few days. The results of the renin and aldosterone were received from the reference laboratory several weeks later
- Plasma renin activity (recumbent) 0.27 ng/h/mL (0.9 pmol/h/mL)
- Aldosterone (recumbent) 45 pg/mL (125 pmol/L)
- These were considered to be inappropriately low for severe hyperkalemia

PATHOPHYSIOLOGY
- The potential for heparin to induce hypoaldosteronism has been recognized for some years. It is a specific, predictable, and reversible effect independent of the route of administration or anticoagulant effect. Hyperkalemia may be induced in as many as 7% of patients with doses as low as 5000 units/d
- Heparin is believed to inhibit the zona glomerulosa (ZG) via angiotensin II receptors within a few days. Narrowing of the ZG has been reported with prolonged use. Patients at highest risk of hyperkalemia are those with already impaired renin-angiotensin-aldosterone axes, such as people with diabetes and patients on angiotensin-converting enzyme inhibitors and nonsteroidal anti-inflammatory drugs
- As aldosterone falls, a secondary increase in plasma renin activity is expected, but was not seen with this patient. This result suggested impaired renin production, perhaps due to nephropathy, although he did not have diabetes, and urea and creatinine were normal after rehydration

FURTHER PROGRESS
- He was later seen by a neurologist who suspected the mitochondrial encephalomyopathy, lactic acidosis, and strokelike episodes (MELAS) syndrome. On investigation, raised serum and cerebrospinal fluid (CSF) lactate were found. Genetic analysis showed a mitochondrial DNA point mutation at nucleotide 3243

FINAL DIAGNOSIS: MELAS syndrome

KEY FEATURES
- MELAS syndrome was first described in 1984. It is a mitochondrial cytopathy inherited through the maternal line, most commonly due to a point mutation A→G at nucleotide 3243
- The defining clinical features are encephalopathy, raised plasma and CSF lactate, and strokes, although these may not all be present at the same time, and patients may not present until middle age. A history of deafness is common. Because it is a cytopathy, many organ systems may be affected and the presenting features are extremely variable. A presentation mimicking viral encephalitis has been reported, as has ischemic colitis. Nephropathy similar to that seen in diabetes has also been described, perhaps explaining this patient's inadequate renin response and his unusually severe hyperkalemia
- It has been postulated that mitochondrial cytopathies present only under metabolic stress, such as febrile illness, when energy requirements exceed supply. Affected organs are damaged by relative cellular adenosine triphosphate deficiency, being pathologically indistinguishable from ischemia
- There is unfortunately no treatment available

POINTS TO REMEMBER
- Recognize the causes of, investigation of, and need for immediate management of hyperkalemia
- Recognize and anticipate the potential for heparin to cause hypoaldosteronism
- Consider MELAS as a cause of recurrent strokes when classical risk factors are absent and when other organ systems are affected

REFERENCES
1. Ostar JR, Singer I, Fishman LM. Heparin-induced aldosterone suppression and hyperkalaemia. Am J Med 1995;98:575–86.
2. O'Kelly R, Magee F, McKenna TJ. Routine heparin therapy inhibits adrenal aldosterone production. J Clin Endocrinol Metab 1983;56:108–12.
3. Pavlakis SG, Phillips PC, DiMauro S, De Vivo DC, Rowland LP. Mitochondrial myopathy, encephalopathy, lactic acidosis and stroke-like episodes: a distinctive clinical syndrome. Ann Neurol 1984;16:481–8.
4. Sharfstein SR, Gordon MF, Libman RB, Malkin ES. Adult-onset MELAS presenting as herpes encephalitis. Arch Neurol 1999;56:241–3.
5. Hess J, Burkhard P, Morris M, Lalioti M, Myers P, Hadengue A. Ischaemic colitis due to mitochondrial cytopathy. Lancet 1995;346:189–90.

THIS CASE WAS PRESENTED BY DR. PAUL MASTERS, CHESTERFIELD ROYAL HOSPITAL, CHESTERFIELD, UK
ACKNOWLEDGMENTS: DR. JOHN T. BOURNE AND DR. MARIOS HADJIVASSILIOU

29. A case of abdominal pain and intermittent diarrhea

PRESENTATION

HISTORY OF PRESENTING COMPLAINT
- A 55-y-old woman was admitted to the emergency department with abdominal pain and hematemesis
- The pain was severe, sharp, and located in the lower abdomen. It radiated to the upper abdomen

DIRECT QUESTIONING
- Loss of appetite
- Weight loss [22 pounds (10 kg) in 5 wk]
- Pale, soft stools
- Drug therapy: Hormone replacement and Motilium (domperidone)

PAST MEDICAL HISTORY
- Three-year history of intermittent diarrhea that had become worse in the past 3 mo

SOCIAL HISTORY
- Nothing of note

FAMILY HISTORY
- Father died aged 52 y from myocardial infarction
- Brother died aged 27 y from a pituitary tumor
- Grandmother and cousin had "problems with their hormones"

ON EXAMINATION
- Patient looked well
- Blood pressure 160/90 mmHg
- Pulse rate 76 beats/min

WHAT IS YOUR PROVISIONAL DIAGNOSIS?

WHAT INVESTIGATIONS WOULD YOU REQUEST?

PROVISIONAL DIAGNOSES
- Peptic ulceration
- Irritable bowel syndrome
- Malabsorption
- Gastrointestinal infection

INITIAL INVESTIGATIONS

Serum		
Sodium	135 mEq/L	(135 mmol/L)
Potassium	3.8m Eq/L	(3.8 mmol/L)
Urea	14 mg/dL	(5.1 mmol/L)
Creatinine	1.1 mg/dL	(95 µmol/L)
Glucose	110 mg/dL	(6.1 mmol/L)
Bilirubin	0.7 mg/dL	(12 µmol/L)
Total protein	6.8 g/dL	(68 g/L)
Albumin	3.7 g/dL	(37 g/L)
Alkaline phosphatase	55 IU/L	(55 IU/L)
Alanine aminotransferase	32 IU/L	(32 IU/L)
Amylase	49 IU/L	(49 IU/L)
Full blood count	Normal	
Mean corpuscular volume	88 µm^3	(88 fl)
Stool culture	Negative	

HAVE YOU MADE ANY CHANGES TO YOUR PROVISIONAL DIAGNOSIS?

WHAT FURTHER INVESTIGATIONS WOULD YOU REQUEST?

A CASE OF ABDOMINAL PAIN AND INTERMITTENT DIARRHEA

WORKING DIAGNOSES
- Peptic ulceration
- Irritable bowel disease
- Malabsorption

FURTHER INVESTIGATIONS

Biochemistry
    Calcium   10.7 mg/dL   (2.68 mmol/L)
    Corrected calcium   12.1 mg/dL   (3.02 mmol/L)

Imaging
    Barium follow through   Consistent with irritable bowel disease or celiac disease
    Computed tomography scan
        Pituitary   No abnormality detected
        Pancreas   Soft tissue mass
    Sigmoidoscopy   No abnormality detected
    Intestinal biopsy   Ulcerative mucosa with excess folding.
        Consistent with celiac, Zollinger-Ellison, or Whipple's disease

- Extra tests at the instigation of the laboratory (abbreviations: GnRH, gonadotropin hormone–releasing hormone; TRH, thyrotropin-releasing hormone; TSH, thyroid-stimulating hormone; LH, luteinizing hormone; FSH, follicle-stimulating hormone; GH, growth hormone)

PITUITARY FUNCTION TESTS

GnRH/TRH TEST	0 MIN	20 MIN	60 MIN
TSH (μU/mL, mU/L)	1.1	15.6	8.1
LH (mU/mL, U/L)	26	>50	>50
FSH (mU/mL, U/L)	32	>50	>50

Normal response

INSULIN TOLERANCE TEST	0 MIN	30 MIN	60 MIN	90 MIN	120 MIN
Cortisol [μg/dL (nmol/L)]	7 (198)	8 (229)	26 (706)	27 (747)	21 (577)
GH (ng/mL, μg/L)	<1	43	88	55	24

Normal response: GH should rise >20
Cortisol should rise above 18.1 (500) with an increase of at least 7.2 (200)

Serum
    Gastrin   560 pg/mL   (560 ng/L)
    Parathyroid hormone   140 pg/mL   (140 ng/L)
    Total thyroxine   10.5 μg/dL   (135 nmol/L)
    Estradiol   51 pg/mL   (189 pmol/L)
    Prolactin   43 ng/mL   (43 μg/L)

- The report included a comment suggesting possible multiple endocrine neoplasia type 1 (MEN type 1)

FINAL DIAGNOSIS: MEN type 1
- Primary hyperparathyroidism: Raised calcium and inappropriately elevated parathyroid hormone
- Pancreatic mass: Peptic ulceration secondary to hypergastrinemia
- Anterior pituitary: Hyperprolactinemia, may be iatrogenic due to domperidone or due to anterior pituitary lesion due to lactotroph

MANAGEMENT
- Commenced treatment with omeprazole (for peptic ulceration)

PROGRESS OVER THE NEXT 3 YEARS
- Weight gain
- Hypertension
- Hypoglycemic attacks

Glucose (fasting)	34 mg/dL	(1.9 mmol/L)
Insulin	55 µU/mL	(380 pmol/L)

FURTHER TREATMENT
- Partial parathyroidectomy
- Vitamin D and calcium
- Octreotide (for hypoglycemia)
- Family referred for genetic assessment
- Patient currently well

KEY FEATURES
- MEN type 1 is a rare disease with an incidence of ~1 in 10,000. It is a familial disorder with an autosomal dominant pattern of inheritance, the affected gene being located on chromosome 11q. Clinical manifestation is usually in the third to fourth decade of life
- The parathyroid, pancreatic islet cells, and pituitary are primarily affected, although the presentation varies in individuals, contributing to underdiagnosis of the disease
- Primary hyperparathyroidism is frequently the first indication of the disease, being present in ~95% of cases. MEN type 1 is not a rare cause of hyperparathyroidism, constituting ~10% of cases
- History and clinical examination should focus on hypercalcemia, peptic ulceration, hypoglycemia, galactorrhea-amenorrhea, Cushing's disease, hypopituitarism, and acromegaly
- The measurement of calcium, glucose, prolactin, and gastrin only may be sufficient in asymptomatic patients with a family history of MEN type 1. Various scanning techniques are used in the location of gastrointestinal tumors. A spectrum of histological features from hyperplasia through multifocal adenomatosis to overt neoplasia may be seen
- Surgery is the treatment of choice for hyperparathyroidism, whereas pharmacological control is indicated for pancreatic tumors
- Once a diagnosis is made, yearly biochemical testing is recommended, and because of MEN type 1's high penetrance, it is usual to screen family members for the disease
- Approximately 50% of MEN type 1 patients die of their disease, as a result of endocrine pancreatic tumors rather than peptic ulceration as the main cause of death

POINTS TO REMEMBER
- To know how to diagnose MEN type 1
- To recognize that the incidence of MEN type 1 is much increased in patients with primary hyperparathyroidism or Zollinger-Ellison syndrome

- To recognize the value of common tests such as calcium in making complex diagnoses
- To have an increased understanding of MEN
- To recognize the role of other specialties in making the diagnosis of MEN type 1
- Domperidone causes hyperprolactinemia

REFERENCES

1. Bone HG. Diagnosis of multiglandular endocrine neoplasias. Clin Chem 1990;36:711–8.
2. Gardner DG. Recent advances in multiple endocrine neoplasia syndromes. Adv Intern Med 1997;42:597–627.
3. Komminoth P, Heitz PU, Kloppel G. Pathology of MEN1: morphology, clinicopathologic correlations and tumor development. J Intern Med 1998;243:455–64.
4. Jialal I, Winter WE, Chan DW. Handbook of diagnostic endocrinology. Washington, DC: AACC Press, 1999;216–23.
5. Dawson DB, Jialal I. Multiple endocrine neoplasia. Diagn Endocrinol Metab 1996;14:579–92.

THIS CASE WAS PRESENTED BY MR. LAURENCE ROBINSON, SOUTH MANCHESTER UNIVERSITY HOSPITAL, MANCHESTER, UK

30. Abdominal pain and vomiting in a 16-y-old girl

PRESENTATION

HISTORY OF PRESENTING COMPLAINT
- A 16-y-old West Indian girl was referred to the surgeons with a week's history of severe abdominal pain and vomiting
- She was unable to eat or drink
- She was on the oral contraceptive pill, and her period had just started

ON EXAMINATION
- She was apyrexial with a pulse of 75 beats/min
- Her abdomen was diffusely tender with no rebound tenderness
- On rectal examination she was tender on the left

INITIAL INVESTIGATIONS
- Urine testing: Glucose +, blood ++, protein +
- Urine microscopy: Confirmed erythrocytes were present with no casts
- Full blood count: Hemoglobin 17.1 g/dL (171 g/L), leukocyte count $17.1 \times 10^3/\mu L$ ($17.1 \times 10^9/L$), and platelets $380 \times 10^3/\mu L$ ($380 \times 10^9/L$)
- She was sickle-cell positive

WHAT ARE YOUR PROVISIONAL DIAGNOSES?

WHAT INVESTIGATIONS ARE INDICATED IN THIS PATIENT?

PROVISIONAL DIAGNOSES
- Urinary tract infection
- Cholecystitis
- Acute pelvic inflammatory disease
- Pancreatitis
- Sickle-cell crisis

FURTHER INVESTIGATIONS
- Biochemical profile, serum and urine amylase
- Abdominal ultrasound scan
- Medical opinion

INITIAL TREATMENT
- Intravenous fluids and prochlorperazine and pethidine for analgesia

MEDICAL ASSESSMENT
- The medical team noted that she had developed a maculopapular rash and their differential diagnosis included porphyria and Henoch-Schönlein purpura in addition to the surgical list

RESULTS
- Ultrasound: Thickened gall bladder with no calculi and normal liver. Prominence of soft tissues posterior to the head of the pancreas. Free fluid noted in the hepatorenal fossa and between the spleen and kidney. The kidneys were normal

BIOCHEMISTRY
Serum

Urea and electrolytes	Normal	
Albumin	2.7 g/dL	(27 g/L)
Bilirubin	1.0 mg/dL	(17 µmol/L)
Amylase	330 IU/L	(330 IU/L)
Phosphate	1.9 mg/dL	(0.62 mmol/L)
Urine amylase	1360 IU/L	(1360 IU/L)
Total protein	5.5 g/dL	(55 g/L)
Alkaline phosphatase	33 IU/L	(33 IU/L)
Alanine aminotransferase	8 IU/L	(8 IU/L)
Calcium	7.1 mg/dL	(1.78 mmol/L)

PROGRESS
- A working diagnosis of resolving pancreatitis was made, and she slowly recovered and was discharged on day 12. Just before discharge her hemoglobin was noted to have fallen to 9.8 g/dL (98 g/L) with an erythrocyte sedimentation rate of 100 mm/h. The results of her final liver function tests were delayed until after her discharge because the sample was somewhat lipemic. A follow up sample was requested for lipid analysis.

RESULTS
Cholesterol	151 mg/dL	(3.9 mmol/L)
Triglycerides	4460 mg/dL	(50.4 mmol/L)

WHAT IS YOUR DIAGNOSIS?

WHAT FURTHER INFORMATION WOULD BE USEFUL?

DIAGNOSIS
- Pancreatitis, secondary to hypertriglyceridemia

MANAGEMENT
- She was referred to the lipid clinic with the above diagnosis
- At the clinic her rash was noted to be prominent but not typical of hypertriglyceridemia
- She had no family history of hyperlipidemia

FURTHER INVESTIGATIONS
- Antinuclear factor of 1 in 100 (possible interference from lipemia)

TREATMENT
- Acipimox (250 mg twice a day)

FURTHER PROGRESS
- Monitored over the next few months, her triglycerides fell to between 180 and 885 mg/dL (2.0 and 10.0 mmol/L)

NEW PRESENTATION
- Three years later she presented to her general practitioner with nausea, decreased appetite, loose stools, and breathlessness
- He measured her hemoglobin and found it to be 4 g/dL (40 g/L)
- On examination she was noted to have a bounding pulse of 120/min, pyrexia, jaundice, splenic enlargement, and vasculitic changes in her fundi

INVESTIGATIONS
- A full blood count showed a raised reticulocyte count [$253 \times 10^3/\mu L$ ($253 \times 10^9/L$)] and a positive Coombs' test

BIOCHEMISTRY
- Normal apart from raised bilirubin of 1.5 mg/dL (25 μmol/L) and globulins of 5.7 g/dL (57 g/L)
- Her antinuclear factor was positive at 1 in 1000 with double-stranded DNA antibodies present

---

WHAT DIAGNOSIS CAN TIE THESE RESULTS TOGETHER?

FINAL DIAGNOSIS: Acute hemolytic anemia secondary to systemic lupus erythematosus (SLE)

PROGRESS
- She made a good recovery on high-dose steroids, which were slowly reduced. The steroids also abolished any need for lipid-lowering therapy to normalize her triglycerides
- It was assumed she had antibodies to lipoprotein lipase that were responsible for the hypertriglyceridemia
- The nature of her original illness was never satisfactorily explained. None of her original samples were lipemic, and her amylase was never more than twice the upper limit of the reference range
- She may have had an intra-abdominal vasculitic episode from her developing SLE

KEY FEATURES
PATHOPHYSIOLOGY
- Autoimmune hyperlipidemia is well described in myeloma. It has been linked to antibodies of immunoglobulin A, G, and M classes. In most cases, the antibodies bind to lipoproteins and alter their metabolism
- A few cases, which have included two cases of SLE with hypertriglyceridemia and one of hyperchylomicronemia during pregnancy, have been described in which antibodies to enzymes necessary for normal lipoprotein lipolysis have been present
- In most of the cases described, the antibodies appear to be directed against heparin
- Treatment is limited to treatment of the primary disease

POINTS TO REMEMBER
- Abdominal pain is a common presenting condition
- It is often easy to diagnose but occasionally causes diagnostic difficulty
- Metabolic problems can cause abdominal pain, such as porphyria, diabetic ketoacidosis, and hyperlipidemia
- Autoimmune disease can cause an array of complications in different organ systems
- Antibodies can be formed against a variety of body constituents

REFERENCES
1. Beaumont JL, Beaumont V. Dyslipoproteinemias and auto immunity. In: Fruchart JC, Shepherd J, eds. Human plasma lipoproteins. Berlin: de Gruyter, 1989:281–307.
2. Yoshimura T, Ito M, Sakoda Y, Kobori S, Okamura H. Rare cases of autoimmune hyperchylomicronemia during pregnancy. Eur J Obstet Gynecol Reprod Biol 1998;761:49–51.

THIS CASE WAS PRESENTED BY DR. TREVOR GRAY, NORTHERN GENERAL HOSPITAL, SHEFFIELD, UK

31. Of course I am taking the tablets, doctor!

PRESENTATION

HISTORY OF PRESENTING COMPLAINT
- A 44-y-old woman, who had been diagnosed as having autoimmune hypothyroidism 4 y previously, was referred to the endocrine clinic because her serum thyroid-stimulating hormone (TSH) concentration remained persistently elevated on replacement therapy of 150 µg of thyroxine (T_4) per day
- She had presented to her primary-care physician in 1991 complaining of excessive tiredness with an elevated TSH of 6 µU/mL (6 mU/L) and a total T_4 of 6.4 µg/dL (83 nmol/L). She had positive thyroid microsomal and thyroglobulin autoantibody titers. There was a family history of autoimmune disease, with her late father having had thyrotoxicosis and type 1 diabetes
- She was commenced on T_4 by her primary-care physician and initially became symptom-free on 150 µg of T_4 per day. However, her TSH remained high [14 µU/mL (14 mU/L)] in spite of an elevated free T_4 (FT_4) at 2.5 ng/dL (32 pmol/L). She became symptomatic again, complaining of lethargy and poor concentration

ON EXAMINATION AND QUESTIONING
- At the endocrine clinic, the patient denied poor adherence to her medication, and the endocrinologist increased her T_4 to 175 mg daily
- On this dose she had some clinical features of thyrotoxicosis with a fine tremor, tachycardia, and mild proximal muscle weakness
- There was no goiter and no visual field defect was detected
- However TSH was elevated at 11 µU/mL (11 mU/L), with a FT_4 of 3.7 ng/dL (47 pmol/L) and free triiodothyronine (FT_3) of 649 pg/dL (10 pmol/L)
- These abnormal results were confirmed using an alternative analytical system

---

WHAT IS YOUR PROVISIONAL DIAGNOSIS?

WHAT FURTHER INVESTIGATIONS WOULD YOU REQUEST?

PROVISIONAL DIAGNOSIS
- Poor compliance should still be considered
- Assay interference is unlikely, but should still be considered
- Primary hypothyroidism with end-organ resistance
- Primary hypothyroidism with a concomitant TSH-secreting tumor

FURTHER TESTS AND INFORMATION
- On further questioning the patient insisted she was taking T_4 diligently
- Treatment to remove heterophilic antibodies from serum produced no change in any of the thyroid function tests. On dilution TSH values showed good linearity
- Primary hypothyroidism coexisting with end-organ resistance is extremely rare
- Primary hypothyroidism with a TSH-secreting tumor has only been described twice in humans

---

WHAT FURTHER INVESTIGATIONS WOULD YOU REQUEST?

HAVE YOU MADE ANY CHANGES TO YOUR PROVISIONAL DIAGNOSIS?

WHAT IS YOUR DIAGNOSIS?FURTHER INVESTIGATIONS

- Prolactin was elevated on two occasions at 43 ng/mL (860 mU/L) and 74 ng/mL (1500 mU/L)
- The α-subunit of glycoprotein hormone was increased at 7.6 ng/mL (7.6 μg/L)
- Magnetic resonance imaging (MRI) revealed a macroadenoma of the right lobe of the pituitary gland, measuring 1.4 cm × 1.3 cm × 1.2 cm, with extension into the suprasellar cistern but no compression of the optic chiasm
- A combined thyrotropin-releasing hormone (TRH) and gonadotropin hormone–releasing hormone (GnRH) test was performed. There was a moderate increment in the prolactin concentration but a failure of the TSH to rise significantly in response to TRH. A normal response to GnRH was obtained

FINAL DIAGNOSIS: The biochemistry and MRI scan are consistent with a TSH-secreting tumor and concurrent autoimmune primary hypothyroidism

MANAGEMENT AND FOLLOW-UP

- The tumor was removed trans-sphenoidally. Histological examination showed the presence of a dual population of tumor cells. Large polygonal cells stained positive for TSH. There was more widespread positivity for α-subunit, and positive staining for prolactin was noted in a subgroup of smaller cells
- Postoperative MRI scan showed no obvious residual tumor. One month postoperation on 175 μg of T_4 per day, her TSH was normal at 2.4 μU/mL (2.4 mU/L). α-Subunit and prolactin were also normal. Tests of anterior pituitary function performed the following month showed no evidence of deficiency of any pituitary hormone. Six months postoperatively, her FT_4 was 2.2 ng/dL (28 pmol/L) but the TSH value had crept up again to 8.5 μU/mL (8.5 mU/L). Her serum prolactin concentration was also slightly elevated at 41 ng/mL (821 mU/L). She was therefore referred for a course of external pituitary irradiation

KEY FEATURES

- TSH-secreting pituitary adenomas are extremely rare, with ~280 being described (1). Most of these adenomas secrete TSH either alone or in combination with α-subunit. However, co-secretion of other pituitary hormones is less well recognized, with only 11% of adenomas secreting both TSH and prolactin (2)
- This case is very unusual in that autoimmune hypothyroidism was co-existing with the pituitary tumor. In mice, gross adenomatous change occurs in the pituitary 10 mo after induction of hypothyroidism (3), but such a transition has not been clearly established in humans. It remains unclear how frequently, if at all, autonomous TSH secretion occurs in humans with primary hypothyroidism (4)
- In our patient, at the time of initial presentation, although TT_4 remained within the low-normal range, the clinical and biochemical features along with the strongly positive thyroid antibodies were entirely consistent with a diagnosis of mild primary autoimmune hypothyroidism. Thyrotroph hyperplasia secondary to primary hypothyroidism regresses with T_4 replacement. Despite adequate replacement of T_4, however, the TSH failed to suppress, thereby raising the suspicion of autonomous TSH secretion. This case clearly illustrates the difficulty in diagnosing either condition in the unlikely event of their co-existing

- Hyperprolactinemia can occur in patients with primary hypothyroidism (because TRH is a stimulating lactotroph); however, in our patient, typical prolactin-secreting tumor cells were clearly demonstrated histologically
- Trans-sphenoidal surgery remains the favored method of treatment for TSHomas. Following this, radiotherapy is conventionally used to treat any residual tissue. Somatostatin analogs (octreotide, lanreotide) may be used as a preoperative debulking measure or as an adjunct to radiotherapy (5)

POINTS TO REMEMBER

- Thyroxine therapy at doses that increase FT_4 and FT_3 should usually suppress serum TSH to low or undetectable concentrations
- Poor compliance and assay interference are considered to be the most common causes of failure to suppress TSH when FT_3 and FT_4 are elevated
- If assay interference and poor compliance can be excluded, primary hypothyroidism with thyroid hormone resistance or TSH-secreting tumor should be considered in patients who fail to suppress TSH when FT_4 and FT_3 are elevated
- The diagnosis of a TSH-secreting adenoma requires a high index of clinical suspicion in all cases, but especially if there is coincidental autoimmune hypothyroidism (albeit this is very rare)

REFERENCES

1. Beck-Peckoz P, Brucker-Davis F, Persani L, Smallridge RC, Weintraub BD. Thyrotropin-secreting pituitary tumors. Endocr Rev 1996;17:610–38.
2. Pioro EP, Scheitauer BW, Laws ER, Randall RV, Kovacs KT, Horvath E. Combined lactotroph and thyrotroph hyperplasia simulating prolactin-secreting adenoma in longstanding primary hypothyroidism. Surg Neurol 1988;29:218–26.
3. Purves HD, Griesbach WE. Changes in basophil cells of rat pituitary after thyroidectomy. J Endocrinol 1956;13:365–75.
4. Scheitauer BW, Kovacs KT, Randall RV, Ryan N. Pituitary gland in hypothyroidism. Histologic and immunocytologic study. Arch Pathol Lab Med 1985;109:499–504.
5. McDermott MT, Ridgway EC. Central hyperthyroidism. Endocrinol Metab Clin North Am 1998;27187–203.

THIS CASE WAS PRESENTED BY DR. GEOFFREY J. BECKETT, DR. JYOTHI M. IDICULLA, AND DR. ALAN W. PATRICK OF THE ROYAL INFIRMARY, EDINBURGH, UK

32. A tale of two cations

PRESENTATION

HISTORY OF PRESENTING COMPLAINT
- A 55-y-old woman of Middle Eastern origin was admitted for assessment of suitability for liver transplantation
- Clinically she had signs of chronic liver disease
- The patient was known to be polymerase chain reaction (PCR)–positive for hepatitis C RNA and to have diabetes mellitus, treated with insulin

SOCIAL HISTORY
- The patient was a housewife. She had 13 children and a large extended family
- She did not smoke, nor did she drink alcohol

FAMILY HISTORY
- There was no family history of liver disease or any inherited disorder

INITIAL INVESTIGATIONS

The laboratory assessment revealed the following:

Serum

Bilirubin	2.6 mg/dL	(44 µmol/L)
Aspartate aminotransferase	227 IU/L	(227 IU/L)
Alanine aminotransferase	214 IU/L	(214 IU/L)
Alkaline phosphatase	634 IU/L	(634 IU/L)
Albumin	3.2 g/dL	(32 g/L)
C-reactive protein	3.2 mg/dL	(32 mg/L)
Immunoglobulin G	3.58 g/dL	(35.81 g/L)
Immunoglobulin A	0.96 g/dL	(9.64 g/L)
Immunoglobulin M	1.36 g/dL	(13.64 g/L)
D-Dimer	1000 ng/mL	(1000 µg/L)
Ceruloplasmin	<6 mg/dL	(<0.06 g/L)
α_1-Antitrypsin	155 mg/dL	(1.55 g/L)

Blood

Hemoglobin A_{1C}	10.5%
Hepatitis C antibody	Positive
Prothrombin time	24.05 s (control 13.5 s)

WHAT CAUSES OF LIVER DISEASE SHOULD BE CONSIDERED FURTHER?

WHAT FURTHER INVESTIGATIONS SHOULD BE PERFORMED?

PROVISIONAL DIAGNOSES
- Chronic hepatitis C infection
- Wilson's disease

FURTHER INVESTIGATIONS

Serum copper	120 µg/dL	(18.8 µmol/L)
24-h urinary copper excretion	64 µg/24 h	(1.0 µmol/24 h)
Serum ferroxidase activity	437 U/L (mean of control group values 686)	
Serum iron	49 µg/dL	(8.8 µmol/L)
Serum transferrin	159 mg/dL	(1.59 g/L)
Serum ferritin	4.5 µg/dL	(45 µg/L)
Analytical interference by immunoglobulins	Excluded heterophilic antibody	
Ocular examination	Kayser-Fleischer rings absent	
Selective magnetic resonance imaging (MRI)		
Liver, pancreas	No evidence of deposition of paramagnetic material	
Head	Evidence of deposition of a paramagnetic material within the lentiform nuclei and heads of caudal nuclei	

ARE THESE FINDINGS CONSISTENT WITH WILSON'S DISEASE?

WHAT ARE THE POSSIBLE CAUSES OF LOW SERUM CERULOPLASMIN?

WHAT OTHER PROPERTY OF CERULOPLASMIN CAN BE USED TO QUANTIFY IT IN SERUM?

DIFFERENTIAL DIAGNOSIS

- The findings are not consistent with Wilson's disease, in which, in addition to the low serum ceruloplasmin concentration, serum copper is usually low and urinary copper excretion is >100 mg/24 h (>1.6 µmol/24 h)
- Kayser-Fleischer rings caused by deposition of copper in Descemet's membrane around the periphery of the cornea, are a characteristic clinical finding

Possible causes of low or undetectable concentrations of ceruloplasmin in serum
- Wilson's disease
- Hereditary aceruloplasminemia
- Chronic liver disease
- Analytical interference

MANAGEMENT

- The patient's diabetes was brought under control, and she underwent liver transplantation
- Her liver function improved sufficiently for her to leave hospital and return home, on azathioprine and cyclosporine immunosuppression

HISTOLOGICAL DIAGNOSIS
- Histology of the resected liver showed an established cirrhosis, consistent with end-stage hepatitis C infection
- There was no evidence of excessive hepatic deposition of copper or iron, using special staining techniques

FURTHER LABORATORY INVESTIGATIONS
- Serum ceruloplasmin was measured on several occasions in the weeks after transplantation. On each occasion, the concentration was within the reference range 28, 31, and 28 mg/dL (0.28, 0.31, and 0.28 g/L)

FINAL DIAGNOSIS: Chronic hepatitis C infection leading to reduced hepatic synthetic capacity for ceruloplasmin

- As well as its function as a carrier protein for copper, ceruloplasmin possesses ferroxidase activity, converting Fe(II) to Fe(III) and facilitating the binding of Fe(III) to transferrin for transport
- In the rare autosomal recessive condition hereditary aceruloplasminemia, transport of iron is severely impaired, resulting in iron deposition in the tissues and low serum iron and high serum ferritin concentrations
- Iron deposition in the pancreas and the liver can lead to diabetes mellitus and chronic liver disease
- The results of MRI and histological examination of the liver are not consistent with life-long copper or iron deposition in the tissues, excluding the diagnoses of Wilson's disease and homozygosity or heterozygosity for hereditary aceruloplasminemia, respectively
- The relatively low serum ferroxidase activity is consistent with the low ceruloplasmin concentration being a genuine finding, not an artifact due to analytical interference
- Serial dilution experiments ruled out analytical interference due to the presence of a heterophilic antibody
- The most likely cause of the low serum ceruloplasmin concentration in this patient is reduced synthetic capacity for ceruloplasmin by the diseased liver
- This is consistent with concentrations of the protein returning to normal after transplantation
- The paramagnetic material detected on MRI of the patient's head may have been due to iron

deposition resulting from reduced ferroxidase activity during the course of the hepatitis C infection

POINTS TO REMEMBER
- Caution should be exercised in interpreting low serum ceruloplasmin in the absence of other supporting evidence of Wilson's disease
- Alternative approaches to measurement of plasma constituents should be considered when analytical interference is a possibility
- Reduced synthetic capacity is a common feature of chronic liver disease. In some instances, the synthesis of individual proteins is selectively affected, and this may have an immunological basis

REFERENCES
1. Cauza E, Maier-Dobersberger T, Polli C, Kaserer K, Kramer L, Ferenei P. Screening for Wilson's disease in patients with liver diseases by serum caeruloplasmin. J Hepatol 1997;27:358–62.
2. Logan JI, Harveyson KB, Wisdom GB, Hughes AE, Archbold GPR. Hereditary caeruloplasmin deficiency, dementia and diabetes mellitus. Q J Med 1994;87:663–70.
3. Cox DW. Genes of the copper pathway. Am J Hum Genet 1995;56:828–34.
4. Jones RJ, Lewis SJ, Smith JM, Neuberger J. Undetectable serum caeruloplasmin in a woman with chronic hepatitis C infection. J Hepatol 2000;32:703–4.

THIS CASE WAS PRESENTED BY MS. JANET M. SMITH, UNIVERSITY HOSPITAL BIRMINGHAM NHS TRUST, BIRMINGHAM, UK

33. Hypocalcemia in the intensive-care unit

PRESENTATION

HISTORY OF PRESENTING COMPLAINT
- A 77-y-old woman was admitted to the hospital for weight loss
- She had an abdominal X-ray, which showed no abnormality, but on clinical examination, a thyroid nodule was palpable
- She suffered a respiratory arrest on the ward

PAST MEDICAL HISTORY
- Nothing of note; no history of thyroid disease

ADMISSION TO INTENSIVE-CARE UNIT
- The patient was intubated with difficulty due to clenched teeth and was mechanically ventilated

INITIAL INVESTIGATIONS

	ON ADMISSION		DAY 2	
Sodium	138 mEq/L	(138 mmol/L)	138 mEq/L	(138 mmol/L)
Potassium	2.7 mEq/L	(2.7 mmol/L)	5.0 mEq/L	(5.0 mmol/L)
Urea	18 mg/dL	(6.9 mmol/L)	41 mg/dL	(14.5 mmol/L)
Creatinine	0.7 mg/dL	(66 µmol/L)	1.9 mg/dL	(164 µmol/L)
Calcium, total	9.3 mEq/L	(2.33 mmol/L)	7.9 mEq/L	(1.98 mmol/L)
Calcium, ionized	5.2 mg/dL	(1.30 mmol/L)	3.2 mg/dL	(0.81 mmol/L)
Albumin	3.3 g/dL	(33 g/L)	–	–
Magnesium	2.1 mg/dL	(0.88 mmol/L)	1.1 mg/dL	(0.45 mmol/L)
Phosphate	2.4 mg/dL	(0.78 mmol/L)	–	–

PROGRESS

- On admission, the patient was given intravenous potassium after the receipt of the laboratory results
- On day 2 the patient became hypotensive, and noradrenalin was started. The potassium infusion was stopped later that day

---

WHY WAS THE PATIENT HYPOCALCEMIC?

WHY WAS THE PATIENT HYPOMAGNESEMIC?

WHY DID THE PATIENT'S RENAL FUNCTION DETERIORATE?

WHAT IS YOUR PROVISIONAL DIAGNOSIS FOR THESE CHANGES?

WHAT FURTHER INVESTIGATIONS WOULD YOU REQUEST?

INITIAL WORKING DIAGNOSIS
- Acute hypocalcemia and possible hypomagnesemia due to hypoparathyroidism

DAY 3
- Calcium and magnesium infusions were started
- The potassium infusion was stopped
- The patient became hypertensive, and the noradrenalin was stopped. In view of her labile blood pressure, a request was made by the anesthetists for urine catecholamines to exclude a pheochromocytoma

Serum
Sodium	139 mEq/L	(139 mmol/L)
Potassium	4.4 mEq/L	(4.4 mmol/L)
Urea	59 mg/dL	(21.2 mmol/L)
Creatinine	2.6 mg/dL	(228 µmol/L)
Calcium, total	4.9 mg/dL	(1.23 mmol/L)
Calcium, ionized	2.4 mg/dL	(0.61 mmol/L)
Albumin	2.3 g/dL	(23 g/L)
Phosphate	8.9 mg/dL	(2.87 mmol/L)
Parathyroid hormone	265 pg/mL	(28.1 pmol/L)

Urine
Calcium	80 mg/24 h	(2.0 mmol/24 h)
Phosphate	1443 mg/24 h	(46.6 mmol/24 h)

- The parathyroid hormone concentration on the sample from day 2 was 264 pg/mL (28.0 pmol/L)

WHY DID THE PATIENT BECOME HYPERTENSIVE?

WHAT CHANGES, IF ANY, HAVE YOU MADE TO YOUR PROVISIONAL DIAGNOSIS?

FURTHER PROGRESS
DAY 4
- The patient was now normotensive

Sodium	136 mEq/L	(136 mmol/L)
Potassium	4.4 mEq/L	(4.4 mmol/L)
Urea	78 mg/dL	(27.8 mmol/L)
Creatinine	2.9 mg/dL	(255 µmol/L)
Calcium, total	9.2 mg/dL	(2.30 mmol/L)
Albumin	2.8 g/dL	(28 g/L)
Magnesium	3.9 mg/dL	(1.62 mmol/L)
Urine catecholamines	Within normal limits	

DAYS 5–10
- The patient's renal impairment resolved, further calcium and magnesium supplements were not required, and the blood pressure remained stable with no further inotrope requirements
- There was never any significant respiratory dysfunction, and the patient was extubated and discharged from the intensive-care unit on day 10

DAY 10 RESULTS

Sodium	144 mEq/L	(144 mmol/L)
Potassium	4.9 mEq/L	(4.9 mmol/L)
Urea	45 mg/dL	(15.9 mmol/L)
Creatinine	1.1 mg/dL	(96 µmol/L)
Calcium, total	8.4 mg/dL	(2.11 mmol/L)
Calcium, ionized	4.7 mg/dL	(1.18 mmol/L)
Albumin	3.2 g/dL	(32 g/L)

FURTHER INVESTIGATIONS
- Inspection of the drug chart revealed that the potassium infusion on day 1 was given as potassium dihydrogen phosphate (KH_2PO_4) rather than as potassium chloride
- The patient was given nearly 2.5 g (80 mmol) over 28 h. Normal phosphate intake is variable, but said to be between 495 and 1486 mg/24 h (16 and 48 mmol/24 h)

FINAL DIAGNOSIS
- Acute phosphate loading caused hypocalcemia and hypomagnesemia, leading to hypotension due to effects on smooth muscle
- The acute reversible renal impairment was due to nephrocalcinosis (deposition of calcium and phosphate within the kidney)
- The hypertension on day 3 reflected the effect of noradrenalin when the hypocalcemia had been corrected with a calcium infusion

KEY FEATURES
- The appropriately high parathyroid values excluded hypoparathyroidism, and the low renal calcium excretion suggested renal tubular function was normal. The high serum phosphate and large phosphate excretion suggested phosphate loading
- Phosphate loading leads to hypocalcemia and hypomagnesemia and may cause soft tissue calcification
- In intensive-care patients, it is important to compare unexpected laboratory results with treatment regimes

POINTS TO REMEMBER

- Measured total calcium comprises ionized, complexed, and albumin-bound fractions; however, it is the ionized fraction that is physiologically active
- Attempts to correct total calcium for albumin assume normal phosphate and acid-base status and are, at best, approximate
- Patients in intensive care are more likely to have derangements of calcium-binding anions such as phosphate and abnormal acid-base balance. For this reason, measurement of ionized calcium is a better marker of calcium homeostasis than total calcium
- Phosphate loading leads to an increase in calcium and magnesium complexes. Although measurement of total magnesium and calcium will include the complexed fraction, when the calcium phosphate solubility product is exceeded, there is a risk of soft tissue deposition of calcium and possibly magnesium salts, leading to hypocalcemia and hypomagnesemia

REFERENCES

1. Endres DB, Rude RK. Mineral and bone metabolism. In: Burtis CA, Ashwood ER, eds. Tietz textbook of clinical chemistry, 3rd ed. 1999:1395–457.
2. Gray TA, Paterson CR. The clinical value of ionised calcium assays. Ann Clin Biochem 1988;25:210–9.
3. Yeung SJ, McCutcheon IE, Schultz P, Gagel RF. Use of long-term intravenous phosphate infusion in the palliative treatment of tumor-induced osteomalacia. J Clin Endocrinol Metab 2000;85:549–55.

THIS CASE WAS PRESENTED BY DR. PETER GOSLING, UNIVERSITY HOSPITAL BIRMINGHAM NHS TRUST, BIRMINGHAM, UK

34. Tale of the unexpected

PRESENTATION
PAST MEDICAL HISTORY
- A 10-y-old boy was admitted to his local hospital with a 3-wk history of bloody diarrhea
- He also had had a 6-mo history of night sweats, loss of appetite, and weight loss
- He had had successful chemotherapy for Burkitt's lymphoma at 2 y of age

ON EXAMINATION
- Pyrexial
- Moderately dehydrated
- Multiple enlarged lymph nodes
- Large abdominal mass in region of sigmoid colon

EARLY MANAGEMENT AND INVESTIGATIONS
- Started on fluids and antibiotics
- Transferred to a teaching hospital for investigation of abdominal mass
- Computed tomography scan showed large pelvic mass involving sigmoid colon and a provisional diagnosis of lymphoma was made (possible recurrence of Burkitt's or second primary tumor)
- Fine-needle aspirate confirmed recurrence of Burkitt's lymphoma
- Combination chemotherapy started with cyclophosphamide-vincristine- methotrexate
- Renal (including urate) and bone (including magnesium) profiles every 4 h were requested for the first 24 h and every 6 h for the next 48 h

CLINICAL BIOCHEMISTRY
- No significant changes in serum biochemistry were seen across the first 24 h of chemotherapy, and results at 24 h were:

Sodium	137 mEq/L	(137 mmol/L)
Potassium	3.4 mEq/L	(3.4 mmol/L)
Bicarbonate	24 mEq/L	(24 mmol/L)
Urea	4 mg/dL	(1.4 mmol/L)
Creatinine	0.47 mg/dL	(42 µmol/L)
Urate	<0.2 mg/dL	(<0.01 mmol/L)
Calcium	9.0 mg/dL	(2.25 mmol/L)
Phosphate	3.0 mg/dL	(0.97 mmol/L)
Albumin	3.0 g/dL	(30 g/L)
Alkaline phosphatase	119 IU/L	(119 IU/L)
Magnesium	1.8 mg/dL	(0.73 mmol/L)

WHAT IS THE RATIONALE BEHIND THE INTENSIVE BIOCHEMICAL MONITORING OF THIS PATIENT DURING THE EARLY PHASE OF CHEMOTHERAPY?

WHAT MECHANISM MIGHT EXPLAIN THE FINDING OF "UNDETECTABLE" SERUM URATE IN A PATIENT?

RATIONALE BEHIND INTENSE BIOCHEMICAL MONITORING
- Patients receiving chemotherapy for bulky and rapidly growing tumors (such as Burkitt's lymphoma) are at high risk of tumor lysis syndrome, particularly during the first phase
- This complication is characterized biochemically by hyperuricemia, hyperphosphatemia, hyperkalemia, and hypocalcemia
- The rapid increase in serum urate and phosphate concentrations can precipitate acute renal failure in this situation unless appropriate prophylactic measures, including vigorous hydration, are taken

POSSIBLE MECHANISMS FOR A FINDING OF "UNDETECTABLE" SERUM URATE
- Marked hypouricemia is an uncommon biochemical finding. Mechanisms include:
  - Increased renal urate loss associated with:
    - Renal tubular disorders (isolated defect or as part of the Fanconi syndrome)
    - Uricosuric drugs
    - Syndrome of inappropriate antidiuretic hormone secretion
  - Impaired urate synthesis
    - Severe liver disease
    - Reduced xanthine oxidase activity: Inherited deficiency (single defect or as part of molybdenum co-factor deficiency); due to "hypouricemic" therapy with xanthine oxidase inhibitors (such as allopurinol)
  - Analytical interference
    - Negative interference with uricase-peroxidase–based procedures occurs in icterus, supraphysiological concentrations of ascorbic acid, and grossly elevated xanthine concentrations associated with use of allopurinol during chemotherapy

FURTHER BIOCHEMICAL AND MANAGEMENT DATA
- No biochemical evidence of tumor-lysis syndrome during chemotherapy
- Urate remained undetectable [<0.2 mg/dL (<0.01 mmol/L)] in the serum for 5 d across the first course of chemotherapy. Concentrations then returned to normal over a further 3 d
- A second course of chemotherapy followed immediately, but serum urate remained normal throughout
- Allopurinol was given prophylactically during both courses of chemotherapy
- The child was discharged home after a total stay of 3 wk with normal renal and bone profiles. The serum urate was 2.2 mg/dL (0.13 mmol/L)

WHAT IS THE MECHANISM FOR THE SERUM URIC ACID FINDINGS IN THIS CASE?

UNDERLYING MECHANISM OF HYPOURICEMIA?
- No analytical explanation for this consistent finding, although negative interference by xanthine was not considered at the time
- Transient nature of the abnormality effectively rules out an inherited defect
- Liver dysfunction was only mild
- Experience suggests that allopurinol alone is unlikely to cause such profound hypouricemia, and its administration was common to both cycles of chemotherapy

FINAL EXPLANATION: Administration of Uricozyme™ (urate oxidase) across first period of chemotherapy
- Uricozyme™ is currently not licensed in the United Kingdom and is available to clinicians on a named-patient basis only
- The preparation is derived from *Aspergillus flavus* and converts uric acid to freely soluble allantoin for ready elimination via the renal route
- Its prophylactic role in chemotherapy-associated hyperuricemia is being assessed in children with high tumor burden, particularly during the first course of chemotherapy
- It is potentially less toxic and more effective than allopurinol in this patient group

POINTS TO REMEMBER
- The importance of a current drug history in the interpretation of unexpected biochemical findings
- The causes of hypouricemia
- The level of biochemical support required to patients at risk of tumor lysis syndrome

REFERENCES
1. Monballyu J, Zachee P, Verberckmoes R. Transient acute renal failure due to tumour lysis induced severe phosphate load in a patient with Burkitt's lymphoma. Clin Nephrol 1984;22:47–50.
2. Dwosh IL, Roncardi DAK, Marliss E, Fox IH. Hypouricaemia in disease: a study of different mechanisms. J Lab Clin Med 1977;90:153–61.
3. Yanasze M, Nakahama H, Mikami H, Fukuhara Y, Orita Y, Yoshikawa H. Prevalence of hypouricaemia in an apparently normal population. Nephron 1988;48:80–3.
4. Crook M. Hyponatraemia and hypouricaemia. Ann Clin Biochem 1993;30:217–8.
5. Leary NO, Pembroke A, Duggan PF. Adapting the uricase/peroxidase procedure for plasma urate to reduce interference due to haemolysis, icterus or lipaemia. Ann Clin Biochem 1992;29:85–9.

THIS CASE WAS PRESENTED BY MR. COLIN T. SAMUELL, UNIVERSITY COLLEGE LONDON HOSPITALS, LONDON

35. What a surprise!

PRESENTATION

HISTORY OF PRESENT COMPLAINT
- A 63-y-old woman was referred by her primary-care physician to a consultant surgeon with a history of rectal bleeding and a tendency to constipation

PAST MEDICAL HISTORY
- Nothing of note

SOCIAL HISTORY
- She was married, was a nonsmoker, and drank alcohol only socially

FAMILY HISTORY
- There was no family history; her parents both died of old age

DRUG HISTORY
- She was on no medication

ON EXAMINATION
- There was no abdominal tenderness; digital rectal examination showed no abnormality

PRELIMINARY INVESTIGATIONS
- Barium enema demonstrated a fungating tumor in the proximal sigmoid colon
- She was subsequently admitted for left hemicolectomy and preoperative investigations were as follows:

    Serum
Corrected calcium	15.2 mg/dL	(3.8 mmol/L)
Alkaline phosphatase	146 IU/L	(146 IU/L)
Creatinine	0.6 mg/dL	(49 µmol/L)

- Patient asymptomatic

WHAT IS YOUR PROVISIONAL DIAGNOSIS?

WHAT INVESTIGATIONS WOULD YOU REQUEST?

PROVISIONAL DIAGNOSIS
- Hypercalcemia, due to:
    Malignancy
    Primary hyperparathyroidism
    Disorders of vitamin D
    Granulomatous disease
    Nonparathyroid endocrine causes, such as thyrotoxicosis
    Drugs, such as thiazide diuretics

INITIAL INVESTIGATIONS
Serum

Corrected calcium	13.9 mg/dL	(3.48 mmol/L)
Phosphate	1.8 mg/dL	(0.57 mmol/L)
Creatinine	0.6 mg/dL	(55 µmol/L)
Sodium	138 mEq/L	(138 mmol/L)
Potassium	40 mEq/L	(4.0 mmol/L)
Bilirubin	0.6 mg/dL	(10 µmol/L)
Aspartate aminotransferase	35 IU/L	(35 IU/L)
Alkaline phosphatase	170 IU/L	(170 IU/L)
Albumin	4.0 g/dL	(40 g/L)

WHAT FURTHER INVESTIGATIONS, IF ANY, WOULD YOU REQUEST?

HAVE YOU MADE ANY CHANGES TO YOUR PROVISIONAL DIAGNOSIS?

WHAT IS YOUR DIAGNOSIS?

WORKING DIAGNOSIS
- Ninety-five percent of cases of hypercalcemia are due either to malignancy or to primary hyperparathyroidism

FURTHER INVESTIGATIONS

Corrected calcium	13.9 mg/dL	(3.48 mmol/L)
Phosphate	1.9 mg/dL	(0.60 mmol/L)
Intact parathyroid hormone (PTH) assay	580 pg/mL	(60.7 pmol/L)
(No interference from PTH-related protein)		

The patient obviously demonstrated inappropriately raised PTH concentration in the face of hypercalcemia
- The concern was whether the hypercalcemia was due to primary hyperparathyroidism and/or malignancy
- The patient could not have tertiary hyperparathyroidism because her renal function was normal and there was no reason to suspect vitamin D deficiency

FINAL DIAGNOSIS: Hypercalcemia due to primary hyperparathyroidism and/or malignancy

MANAGEMENT

In the first instance, the patient was treated with intravenous (i.v.) furosemide, normal saline, and disodium pamidronate for her hypercalcemia. This regimen can cause hypophosphatemia and can take up to 7 d for patients to respond
- One week after admission

Serum corrected calcium	10.4 mg/dL	(2.60 mmol/L)
Phosphate	1.3 mg/dL	(0.41 mmol/L)

At surgery, she had a left hemicolectomy and history showed a Duke's A carcinoma with no metastases
- At discharge

Serum corrected calcium	10.7 mg/dL	(2.67 mmol/L)

- She was seen in the outpatient clinic 2 wk later, and her corrected calcium had gone up to 13.2 mg/dL (3.29 mmol/L) with a phosphate of 2.2 mg/dL (0.71 mmol/L)
- She was treated again with normal saline, disodium pamidronate, and furosemide and was given an urgent referral for parathyroidectomy
- She underwent total parathyroidectomy, thyroidectomy, and thymectomy and on discharge her corrected calcium was still elevated [11.2 mg/dL (2.80 mmol/L)]. She was readmitted for localization through magnetic resonance imaging and angiography of any remaining parathyroid tumor
- A large parathyroid adenoma was identified halfway down the anterior mediastinum. She was readmitted for median sternotomy and parathyroidectomy
- Discharged normocalcemic on alfacalcidol, Calci-chew, and thyroxine

KEY FEATURES

PRESENTATION
- Symptoms of hypercalcemia for primary hyperparathyroidism are more insidious and patients are often asymptomatic. In patients with malignancy and hypercalcemia, the diagnosis is usually at a more advanced stage of the underlying disease
- Symptoms include specifically hypertension, renal involvement, including polyuria, polydipsia, nephrolithiasis, and progressive loss of renal function
- In addition, symptoms may be associated with central nervous system manifestations ranging from mild personality disturbance to severe psychiatric disorders, muscular involvement including proximal muscle weakness, gastrointestinal tract involvement including vague abdominal complaints and constipation, and, with primary hyperparathyroidism, bone pain associated with osteitis fibrosa cystica
- Also, those patients with primary hyperparathyroidism may have hypertension
- Up to 20% of patients with primary hyperparathyroidism have multiple endocrine neoplasia I or IIa (MEN I or IIa)

BIOCHEMISTRY
- Corrected calcium is elevated in both primary hyperparathyroidism and malignancy
- Serum phosphate concentration will be low in patients with primary hyperparathyroidism, due to the effect of PTH on renal phosphate excretion
- Patients with hypercalcemia due to malignancy can often have a normal or raised serum phosphate
- Alkaline phosphatase activity can be elevated in both causes, due to the increase in activity of bone, alkaline phosphatase isoenzymes due to the effect of parathyroid hormone, or alternatively as a result of malignancy
- Raised liver alkaline phosphatase activity may also be seen as a result of metastases in the liver from the initial malignancy

DIAGNOSIS
- Diagnosis is made by PTH concentration being inappropriately elevated for the serum corrected calcium

Etiology of malignancy related hypercalcemia depends on the type of malignancy

CLASSIFICATION OF TUMOR HYPERCALCEMIA
Hematological malignancies
  a) Local bone destruction (osteoclast-activating factor, interleukin 1, tumor necrosis factor, lymphotoxin)
     Multiple myeloma
     Lymphomas
  b) Humoral mediation [1,23-dihydroxycholecalciferol, PTH-related peptide (PTH-rP)]
     Lymphomas
Solid tumors
  a) Local bone destruction (prostaglandin E series)
     Some breast cancers
  b) Humoral medication (PTH-rP, possibly other agents)
     Lung squamous cell
     Kidney
     Urogenital tract
     Other breast cancers

TREATMENT
- Treat the hypercalcemia with i.v. fluid, furosemide, and biphosphonate, either first generation, which is etidronate, or second generation, which is disodium pamidronate
- Other treatments such as calcitonin can be used, but have limited calcium-lowering properties and can produce tachyphylaxis. The steroid glucocorticoids may be used in patients with hypercalcemia due to malignancy
- Treatment of primary hyperparathyroidism is parathyroidectomy, and it has to be remembered that there may be more than one abnormal parathyroid gland and/or the parathyroid tumor may be in an atypical position

Surgery is recommended if patients meet one of the following criteria:
- Serum corrected calcium is >12.0 mg/dL (>3.0 mmol/L)
- Marked hypercalciuria is >36 mg/24 h (>9 mmol/24 h)
- Overt manifestation of primary hyperparathyroidism (nephrolithiasis, osteitis fibrosa cystica, neuromuscular disease)
- Marked reduction in cortical bone density (radial density >2 SD below normal)
- Reduced creatinine clearance in absence of another cause
- Aged <50 y old

POINTS TO REMEMBER
- Primary hyperparathyroidism is common, with a prevalence of between 1 in 200 and 1 in 2000
- The diagnosis is based on the combination of elevated serum calcium and elevated PTH
- Symptoms of hypercalcemia may be rather nonspecific and include thirst, polyuria, fatigue, depression, nausea, and constipation
- Surgery is usually indicated in the patient with symptomatic hypercalcemia
- In the patient with asymptomatic hypercalcemia, the decision regarding surgery should be based on factors such as concentrations of serum calcium, presence of renal stones and/or hypercalciuria, and the measured bone density
- Don't jump to conclusions

REFERENCES
1. Allerheiligen DA, Schoeber J, Houston RE, Mohl VK, Wildman KM. Hyperparathyroidism. Am Fam Physician 1998;57:1795–1802,1807–8.
2. Deftos LJ. Hypercalcaemia. Postgrad Med 1996;100:6:119–26.
3. Silverberg SJ. Diagnosis, natural history and treatment of primary hyperparathyroidism. Cancer Treat Res 1997; 89:163–81.
4. Strewler GJ. The physiology of parathyroid hormone related protein. New Engl J Med 2000; 342:3:177–85.

THIS CASE WAS PRESENTED BY DR. DANIELLE B. FREEDMAN, LUTON AND DUNSTABLE HOSPITAL, LUTON, UK

36. A child with constipation

PRESENTATION

HISTORY OF PRESENTING COMPLAINT
- A 19-mo-old boy was referred to a general pediatrician with a 9-mo history of constipation, poor appetite, and failure to thrive

PAST MEDICAL HISTORY
- Ventose delivery at term
- Normal development until 10 mo of age

FAMILY HISTORY
- No family history of note

MEDICATION
- Lactulose 10 mL twice a day and carbohydrate supplements

ON EXAMINATION
- Very fine fair hair and blue eyes
- Height: 9th centile for age
- Weight: <0.4th centile
- Generalized muscular wasting
- Distended lower abdomen with palpable feces

WHAT ARE YOUR PROVISIONAL DIAGNOSES?

WHAT INVESTIGATIONS WOULD YOU REQUEST?

PROVISIONAL DIAGNOSES
- Cystic fibrosis or celiac disease causing malabsorption
- Renal glomerular or tubular disease causing poor nutrient utilization
- Inborn error of metabolism
- Psychosocial neglect
- Infection

INITIAL INVESTIGATIONS
Serum

Sodium	138 mEq/L	(138 mmol/L)
Potassium	1.8 mEq/L	(1.8 mmol/L)
Urea	13 mg/dL	(4.7 mmol/L)
Calcium	9.6 mg/dL	(2.40 mmol/L)
Phosphate	3.2 mg/dL	(1.04 mmol/L)
Magnesium	1.6 mg/dL	(0.64 mmol/L)
Alkaline phosphatase	398 IU/L	(398 IU/L)
Alanine aminotransferase	13 IU/L	(13 IU/L)
Albumin	4.4 g/dL	(44 g/L)

Blood

Hemoglobin	9.0 g/dL	(90 g/L)
Mean corpuscular volume	74 µm^3	(74 fL)

Feces and urine culture negative

A repeat potassium measurement 8 h later was 2.0 mEq/L (2.0 mmol/L)

MANAGEMENT
- He was started on potassium supplements at a dose of 3 mmol/kg per day

WORKING DIAGNOSIS
- Severe hypokalemia
- Hypomagnesemia
- Hypophosphatemia and anemia

WHAT FURTHER INVESTIGATIONS, IF ANY, WOULD YOU REQUEST?

A CHILD WITH CONSTIPATION

FURTHER INVESTIGATIONS

Serum
- pH^+ 7.35   (H^+ 45 nmol/L)
- pCO_2 32 mmHg   (4.3 kPa)
- Standard bicarbonate   19.5 mEq/L   (19.5 mmol/L)

Plasma
- Renin   >112 pmol/h/mL   (>137 ng/h/mL)
- Aldosterone   >119 pg/mL   (>3300 pmol/L)

Urine
- pH   7.0
- Sodium   39 mEq/L   (39 mmol/L)
- Potassium   27 mEq/L   (27 mmol/L)
- Chloride   23 mEq/L   (23 mmol/L)
- Amino acids   Generalized aminoaciduria
- Protein-creatinine ratio   7766 mg/g   878.5 mg/mmol
- Calcium-creatinine ratio   99 mg/g   <0.28 mmol/mmol

WORKING DIAGNOSIS
- Low plasma potassium with inappropriately elevated urinary potassium, mild metabolic acidosis, and inappropriately alkaline urine suggest a diagnosis of renal tubular acidosis
- The presence of gross aminoaciduria and proteinuria suggests that the tubular acidosis is proximal

FURTHER INVESTIGATIONS
- Slit-lamp examination of the cornea: presence of fine crystals
- Leukocyte cysteine: 2.34 mg/g protein (19.3 µmol/g protein) (grossly elevated)

FINAL DIAGNOSIS: Cystinosis

EVOLUTION
- Ten days after admission the patient developed frequent episodes of vomiting that resulted in severe dehydration, hypocalcemia [5.4 mg/dL (1.36 mmol/L)], and hypomagnesemia [<0.49 mg/dL (<0.2 mmol/L)]

MANAGEMENT
- He was given intravenous fluids (normal saline), intravenous calcium gluconate, magnesium, and antiemetics
- He remained on potassium supplements and was started on phosphate and sodium bicarbonate supplements. 1,25-Dihydroxyvitamin D was also started
- Cysteamine was given to reduce intracellular cystine accumulation

KEY FEATURES

PRESENTATION
- Children with cystinosis appear healthy at birth and present around 6–9 mo of age with nonspecific symptoms such as poor growth or weight loss, vomiting, constipation, poor feeding, thirst, polyuria, and craving for salt
- Affected children may also present with signs of rickets
- The incidence is about 1 case in 200,000 live births

- The pattern of inheritance is autosomal recessive
- Older children or adults may present with proteinuria or features of renal failure
- Treatment with cysteamine delays the progression of renal failure but does not avoid long-term complications (see table)

TABLE
Long-term complications of cystinosis

Endocrine	Hypothyroidism
	Hypogonadotrophic hypogonadism
	Diabetes mellitus
Neurological	Distal myopathy
	Cerebral degeneration (memory loss, ataxia, seizures)
	Pseudobulbar palsy
	Stroke-like episodes
Ophthalmological	Progressive corneal cystine deposition
	Retinopathy
Gastrointestinal	Swallowing difficulties
	Pancreatic dysfunction
Other	Pancytopenia
	Hepatic dysfunction

BIOCHEMISTRY
- Cystinosis is caused by a defective lysosomal cystine transporter resulting in intracellular cystine accumulation that causes cell dysfunction
- The proximal renal tubule is very sensitive to cystine toxicity, resulting in deficient reabsorption of sodium, potassium, bicarbonate, phosphate, calcium, magnesium, proteins, glucose, and amino acids
- Renal loss of sodium and potassium result in hyponatremia and hypokalemia, which can be life-threatening. They also contribute to the impairment of urine concentration ability
- The loss of bicarbonate produces a hyperchloremic metabolic acidosis. The urine is typically alkaline but will acidify in the presence of severe acidosis [bicarbonate <15 mEq/L (15 mmol/L)]
- The urinary loss of calcium and phosphate results in rickets and contributes to the increased risk of developing nephrocalcinosis
- The increased plasma renin activity and aldosterone concentration reflect secondary hyperaldosteronism due to renal loss of sodium and volume contraction
- Bartter's syndrome also presents in childhood with failure to thrive, renal potassium loss, and hyperreninemic hyperaldosteronism, but there is metabolic alkalosis

DIAGNOSIS
- An elevated concentration of leucocyte cystine (50–100 times normal) is diagnostic for cystinosis
- Cystine crystals seen by slit-lamp examination of the cornea are also diagnostic for cystinosis

POINTS TO REMEMBER
- The diagnosis of cystinosis should be suspected in children with a history of polydipsia and/or polyuria and failure to thrive in whom urinalysis shows proteinuria and glycosuria
- Replacement of electrolyte losses is an essential part of treatment

REFERENCES

1. Van't Hoff W. Nephropathic cystinosis—from infancy to childhood. Brit J Renal Med 1997;4:6–8.
2. Niaudet P, Tete MJ, Broyer M. Renal disease in cystinosis. In: Broyer M, ed. Cystinosis. Paris: Elsevier, 1999:36–41.

THIS CASE WAS PRESENTED BY DR. JOSE CABRERA-ABREU, BIRMINGHAM CHILDREN'S HOSPITAL, BIRMINGHAM, UK

37. A woman eating tuna

PRESENTATION

HISTORY OF PRESENTING COMPLAINT
- A 27-y-old female biomedical scientist was eating lunch
  - She felt tingling of lips while eating
  - Within 10 min, she felt her lips swell and felt agitated and hot
  - Within 30 min, her face was red, and she took two chlorpheniramine tablets
  - Within 45 min, she was red all over, had difficulty in breathing, had palpitations, and felt "impending doom"
  - She collapsed while being taken to the hospital by ambulance
- In the hospital she was resuscitated with intravenous fluids, intramuscular epinephrine, intravenous hydrocortisone, and chlorpheniramine
- She felt back pain, and she vomited

ON EXAMINATION
- Her systolic blood pressure was 80 mmHg
- No other abnormalities found

INITIAL INVESTIGATIONS IN EMERGENCY DEPARTMENT

Serum
Sodium	145 mEq/L	(145 mmol/L)
Potassium	4.2 mEq/L	(4.2 mmol/L)
Urea	15 mg/dL	(5.4 mmol/L)
Creatinine	0.9 mg/dL	(78 µmol/L)

Plasma
Glucose	97 mg/dL	(5.4 mmol/L)

WHAT ARE YOUR PROVISIONAL DIAGNOSES?

WHAT FURTHER INFORMATION WOULD YOU LIKE TO KNOW FROM THE PATIENT?

HOW WOULD YOU INVESTIGATE THIS PATIENT?

PROVISIONAL DIAGNOSES
- Anaphylactic shock
- Severe food allergy

FURTHER INFORMATION

DIRECT QUESTIONING
- Food consumed included tuna, cheese, onion bread containing nuts, green salad (olive oil and basil dressing), and fresh orange juice
- She had had no reaction like this previously
- She had consumed similar meals with no ill effect
- She tolerated other fish (including tuna), shellfish, fruit, eggs, and milk
- She had taken aspirin and antibiotics without problem

PAST MEDICAL HISTORY
- She had mild hayfever and eczema as a child, but there had been no recent symptoms

SOCIAL HISTORY
- Worked in a clinical biochemistry laboratory. No exposure to industrial solvents and no latex allergy
- She drank alcohol in moderation and was a nonsmoker

FAMILY HISTORY
- She thought that her mother was sensitive to some anesthetic drugs

CLINICAL PROGRESS

IMMUNOLOGICAL CONSULTATION 1 (1 WK LATER)
- She was negative to all the common skin-prick tests
- It was concluded that she was probably not allergic (nonatopic) to common foods including fish
- Investigations were arranged and she was seen 1 wk later

IMMUNOLOGICAL CONSULTATION 2 (1 WK LATER)
- The results of investigations were:

Total immunoglobulin E (IgE)	150 IU/mL
Allergen-specific IgE	<0.35 IU/mL

  Allergen-specific IgE tests included house dust mites, grass pollens, peanuts, cheeses, mixed nuts, mixed seafoods, and tuna

  A full blood count revealed no eosinophilia
- These results were in line with the clinical observations indicating that the patient had not had an allergic reaction
- At the clinic she was subjected to a prick-prick test with fresh tuna (the food thought most likely to be a possible problem). This was carried out by pricking the fish directly into the patient's forearm under controlled conditions. No reaction was elicited, thereby providing further evidence that the patient was not allergic to tuna

---

WHAT IS THE FINAL DIAGNOSIS?

FINAL DIAGNOSIS
- It was concluded that the patient had presented with pseudoanaphylaxis due to scombroid fish intoxication
- This is a non-IgE severe reaction (but not true anaphylaxis)

FURTHER CLINICAL PROGRESS
- To confirm conclusively that the patient was not allergic to tuna she received an oral challenge using cooked fresh tuna:

0 min	Touched lip	Negative
20 min	Inside lip	Negative
40 min	Chew and spit	Negative
60 min	Chew and swallow	Negative
90 min	Repeat (larger portion)	Negative

- It was concluded that the patient was not allergic to tuna and was discharged from the clinic with the precautionary advice of carrying an Epipen (injectable epinephrine) with her for 1 y

KEY FEATURES OF SCOMBROID FISH TOXICITY
GENERAL
- These fish include oily fish such as mackerel and tuna
- Scombroid fish toxicity is the commonest worldwide cause of fish poisoning

BIOCHEMISTRY
- Scombroid fish are very active swimmers and cover large distances. They require a large capacity for buffering acids formed during metabolism
- Buffering capacity is increased by the dipeptides carnosine and anserine (both of which are 50% histidine). These dipeptides interact with peroxidation products and protect cell membranes

PATHOPHYSIOLOGY
- Spoiled fish decarboxylate histidine to histamine
- It is the ingestion of large amounts of histamine that causes symptoms in affected individuals
- It is not through the activation of mast cells by an immune process
- Histamine is metabolized partly to methylhistamine, and increased concentrations of this metabolite are found in patients suffering from scombroid fish poisoning

POINTS TO REMEMBER
- Good history is vital
- Be alert to certain foods
- Beware peppery, metallic taste of tuna and other such fish, whether fresh or canned
- Know the value of proper investigation
- Know that all reactions to food are not necessarily allergic
- Encourage good communication between the laboratory and the clinic
- Ensure that there is notification of such incidents with the appropriate local environmental authorities

REFERENCES
1. Morrow JD, Margolies GR, Rowland J, Roberts LJ. Evidence that histamine is the causative toxin of scombroid-fish poisoning. N Engl J Med 1991;324:716–20.

2. Hughes JM, Potter ME. Scombroid-fish poisoning: from pathogenesis to prevention. N Engl J Med 1991;324:766–8.
3. Merson MH, Baine WB, Gangarosa EJ, Swanson RC. Scombroid fish poisoning. JAMA 1974;228:1268–9.
4. Kerr GW, Parke TRJ. Scombroid poisoning—a pseudoallergic syndrome. J R Soc Med 1998;91:83–4.
5. Sabroe RA, Black AK. Scombrotoxic fish poisoning. Clin Exp Dermatol 1998;23:258–9.

THIS CASE WAS PRESENTED BY DR. JAMES HOOPER, ROYAL BROMPTON HOSPITAL, LONDON, UK

38. There's a gap: anion, osmolar, or what?

PRESENTATION

HISTORY OF PRESENTING COMPLAINT
- A 37-y-old woman (MH) was found semiconscious and hypothermic (35.4 °C) in her apartment. She was taken to the accident and emergency department of the local hospital

DIRECT QUESTIONING
- No history could be obtained from the patient

PAST MEDICAL HISTORY
- Discussion with her primary-care physician revealed a history of alcohol abuse with depression

ON EXAMINATION
- She was very thin and covered with bruises
- Her blood pressure was 90/50 mmHg, with a pulse rate of 120 beats/min
- She was tachypneic
- Reflexes were brisk and symmetrical, tone was increased all over, and the patient was rigid
- Staff noticed a "ketotic" smell
- While the staff was examining the patient, a 37-y-old man (HH) was admitted into the accident and emergency department. He was found wandering the streets without shoes, confused and hallucinating. He denied drug abuse but admitted alcohol abuse. He had been to a party the night before and was the partner of the woman admitted earlier

---

WHAT IS YOUR PROVISIONAL DIAGNOSIS?

WHAT INVESTIGATIONS WOULD YOU REQUEST?

PROVISIONAL DIAGNOSIS
- Diabetic ketoacidosis
- Alcoholic ketoacidosis
- Drug abuse
- Other poisoning

INITIAL INVESTIGATIONS

Serum

Sodium	136 mEq/L	(136 mmol/L)
Potassium	3.5 mEq/L	(3.5 mmol/L)
Urea	8 mg/dL	(2.7 mmol/L)
Creatinine	1.2 mg/dL	(110 µmol/L)
Chloride	102 mEq/L	(102 mmol/L)
Bicarbonate	10 mEq/L	(10 mmol/L)
Glucose	70 mg/dL	(3.9 mmol/L)
Calcium	8.8 mg/dL	(2.21 mmol/L)
Albumin	2.8 g/dL	(28 g/L)
Osmolality	285 mOsm/kg	(285 mOsm/kg)
Anion gap	27.5 mEq/L	(27.5 mmol/L)
Acetaminophen (paracetamol)	Not detected	
Salicylate	Not detected	

Urine

Ketones (Acetest)	positive	

Other

pH	7.24	(H$^+$ 58 nmol/L)
pCO_2	17 mmHg	(2.26 kPa)
pO_2	177 mmHg	(23.54 kPa)
Calculated bicarbonate	7.1 mEq/L	(7.1 mmol/L)
Base excess	−17.6 mEq/L	(−17.6 mmol/L)
O_2 saturation	99%	

IDENTIFY THE METABOLIC DISTURBANCE.

HAVE YOU MADE ANY CHANGES TO YOUR PROVISIONAL DIAGNOSIS?

WHAT INVESTIGATIONS WOULD YOU REQUEST?

METABOLIC DISTURBANCE
- The patient has a partially compensated metabolic acidosis with an anion gap. There is no osmolar gap

WORKING DIAGNOSIS
- Lactic acidosis
- Alcohol or solvent poisoning

FURTHER INVESTIGATIONS

Lactic acid	14.7 mg/dL	(1.63 mmol/L)
Methanol	Negative	
Ethanol	Negative	
Microscopic examination of urine revealed no crystals		
Isopropanol	Negative	
Serum acetone	36 mg/dL	(62 mmol/L)

The patient HH had similar biochemistry to the above patient, and his acetone was found to be 33 mg/dL (5.7 mmol/L)

FINAL DIAGNOSIS: Acetone poisoning

MANAGEMENT
- MH recovered full consciousness after a couple of hours
- Both patients said that they had not been drinking ethanol or taking drugs
- They did admit to having made up "something" with orange juice at the party
- Because of the high concentration of acetone, they were admitted overnight for observation and supportive care
- Both patients were discharged after 24 h

KEY FEATURES
BIOCHEMICAL INVESTIGATIONS
- Simple, readily available clinical laboratory tests may provide important clues to the diagnosis of poisoning and may guide the investigation towards specific toxicology testing. Essential initial tests are:
  - Serum osmolality and osmolar gap
  - Urea and electrolytes
  - Bicarbonate, chloride, and anion gap
  - Plasma glucose
  - Acetaminophen (paracetamol) and salicylate
  - Blood gases
  - Calcium
  - Liver function tests and plasma ethanol if necessary
- The anion gap is the difference in the sums of the concentrations (in either conventional or SI units) of the principal cations and the principal anions as follows:

Anion gap = $([Na^+] + [K^+]) - ([Cl^-] + [HCO_3^-])$

The units of the derived parameter are mEq/L (mmol/L)

- The osmolar gap is calculated from the difference between the measured osmolality and the expected value (twice the molar concentration of sodium plus that of urea and glucose) as follows:

   Osmolar gap = measured osmolality − (2 × [Na] + [urea] + [glucose])

   Note: The glucose and urea concentrations must be expressed in molar terms (mmol/L)

BIOCHEMISTRY
- Causes of increased anion gap [>15 mEq/L (>15 mmol/L)] are renal failure; diabetic ketoacidosis; alcoholic ketoacidosis, lactic acidosis; and ingestion of substances such as salicylate, methanol, ethylene glycol, and ethanol
- The urine should be checked for calcium oxalate monohydrate crystals 15 h after suspected ingestion of ethylene glycol
- Acetone in plasma can be caused by acetone ingestion, isopropanol ingestion, and acetone sniffing
- Acetone can be found in nail polish remover. Isopropanol can be obtained as rubbing alcohol and is present in some car windshield washes. Ethylene glycol and methanol can also be present in some brands of windshield washes
- Isopropanol is cleared from the plasma more quickly than acetone. It has a plasma half-life ($t_{1/2}$) of 3–6 h and is converted to acetone by alcohol dehydrogenase. Acetone is eliminated slowly ($t_{1/2}$ of 22 h), primarily in alveolar air and urine. There is some conversion to acetate and formate
- The patients had ingested either acetone or isopropanol. Undetectable isopropanol with significant acetone in plasma would be compatible with isopropanol ingestion over 12 h previously

TOXICITY
- Isopropanol toxicity mimics drunkenness and can cause abdominal pain, hypoglycemia, vomiting, dizziness, muscle incoordination, headache, confusion, stupor, and coma. Concentrations >150 mg/dL (>25.0 mmol/L) usually result in death
- Acetone toxicity can cause hypothermia, hypotension, hypoglycemia, renal tubular necrosis, myopathy and rhabdomyolysis, and hemolytic anemia. Toxic concentrations are 20–30 mg/dL (3.44–5.16 mmol/L), and death can occur with concentrations >50 mg/dL (>8.6 mmol/L)

DIAGNOSIS
- A metabolic acidosis with a high anion gap in the absence of diabetic ketoacidosis, renal failure, and lactic acidosis is suspicious of ingestion of alcohols or solvents
- Neither patient had a significant osmolar gap. It must be remembered that solvents have different toxicities and different osmotic activities. Acetone can be toxic and produce only mild increases in osmolality

	TOXIC VALUES	
	CONCENTRATION [mg/dL (mmol/L)]	OSMOLALITY (mOsm/kg)
Ethanol	500 (108.5)	120
Methanol	50 (15.6)	17
Acetone	50 (8.6)	8.5
Ethylene glycol	100 (16.1)	17

TREATMENT
- Maintain the airway and assist ventilation if necessary. Administer supplemental oxygen
- The patient should be assessed for coma, hypotension, and hypoglycemia and treated if they occur
- Symptomatic patients should be admitted and observed for at least 6–12 h
- There is no specific antidote, and ethanol therapy is not recommended
- Hemodialysis effectively removes isopropyl alcohol and acetone but is rarely indicated because most patients can be managed with supportive care alone
- Creatine kinase should be monitored to investigate for possible rhabdomyolysis

POINTS TO REMEMBER
- Know the basic biochemical tests for the investigation of unknown poisoning
- Recognize the differential diagnoses in metabolic acidosis with increased anion gap
- Appreciate the osmolalities of toxic concentrations of alcohols and solvents

REFERENCES
1. Olson KR, ed. Poisoning and drug overdose, 2nd ed. Norwalk, CT: Appleton and Lange, 1994.
2. Natowicz M, Donahue J, Gorman L, et al. Pharmacokinetic analysis of a case of isopropanol intoxication. *Clin Chem* 1985;31:326–8.
3. Pappas AA, Ackerman BH, Olsen KM, Taylor EH. Isopropanol ingestion: a report of six episodes with isopropanol and acetone serum concentration time data. *J Toxicol Clin Toxicol* 1991;29:11–21.
4. Lacouture PG, Wason S, Abrams A, Lovejoy FH. Acute isopropyl alcohol intoxication. Diagnosis and management. *Am J Med* 1983;75:680–6.
5. Jones AW. Elimination half-life of acetone in humans: case reports and review of the literature. *J Anal Toxicol* 2000;24:8–10.
6. Ho MT, Saunders CE, eds. Current emergency diagnosis and treatment, 4th ed. Norwalk, CT: Appleton and Lange, 1995.

THIS CASE WAS PRESENTED BY DR. MARTIN A. MYERS, ROYAL PRESTON HOSPITAL, PRESTON, UK

39. Long walks and dark urine

PRESENTATION

HISTORY OF PRESENTING COMPLAINT
- A 26-y-old tunneling engineer was seen; he had had an uneventful childhood and early adult life
- He was keen on hiking. However as he walked increasingly long distances, often without taking food, he noted that his muscles became unaccustomedly stiff and that he often passed "rich dark urine" (sic)
- Furthermore, he noted that if he partook of alcohol during a hike, his symptoms were aggravated
- He did not feel weak at rest, but increasingly he felt that he fatigued more easily
- After a collapse while walking on vacation in the Grand Canyon that required renal dialysis for acute renal failure secondary to rhabdomyolysis, he was referred for investigation. At the time of referral he was well and there was no other history of note. He did not suffer from seizures. He was on no medication

ON EXAMINATION
- Clinical examination was normal
- There was no hypertrophy, tenderness, or weakness noted in any muscle groups
- He did not fatigue on clinical assessment, and tone was normal
- His cardiovascular examination was unremarkable
- On initial investigation routine hematology and clinical biochemistry including renal, liver, and thyroid functions were all normal. His creatine kinase was also normal at 188 IU/L. An electrocardiogram showed sinus rhythm, a normal axis, and no evidence of hypertrophy

---

WHAT IS YOUR PROVISIONAL DIAGNOSIS?

WHAT INVESTIGATIONS WOULD YOU REQUEST?

PROVISIONAL DIAGNOSES

In the absence of trauma and an ischemic injury or drug abuse, your provisional diagnosis should have included:
- Heat stroke
- Acute inflammatory (polymyositis)
- Viral necrotizing myositis (Epstein-Barr virus, influenza virus, and Coxsackie virus)
- Chronic alcohol abuse
- Hypokalemia
- Hypophosphatemia and metabolic causes

The latter group would include:
- Glycogenoses (such as McArdle's disease)
- Purine metabolic disorders such as myoadenylate deaminase deficiency
- Disorders of oxidative metabolism, notably fatty acid and respiratory chain disorders

INITIAL INVESTIGATIONS

C-reactive protein	Normal	
Erythrocyte sedimentation rate	Normal	
Plasma uric acid	6.2 mg/dL	(0.37 mmol/L)

He was Monospot negative, and titers for influenza virus and Coxsackie virus were not raised
Forearm ischemic exercise test results

	BASELINE VALUES	PEAK VALUES
Lactate	5.3 mg/dL (0.59 mmol/L)	38.9 mg/dL (4.32 mmol/L)
Lactate-pyruvate ratio	11.0	47.0
Ammonia	42 µg/dL (30 µmol/L)	155 µg/dL (111 µmol/L)

Note: Peak values were noted 2–4 min post ischemia. Performed as described in reference 1

Urine organic acids: No evidence of a metabolic disorder

Total plasma carnitine	0.4 mg/dL	(20 µmol/L)
Free carnitine	0.28 mg/dL	(14 µmol/L)
Derived acyl fraction	0.12 mg/dL	(6 µmol/L)

WHAT FURTHER INVESTIGATIONS, IF ANY, WOULD YOU REQUEST?

HAVE YOU MADE ANY CHANGES TO YOUR PROVISIONAL DIAGNOSIS?

WHAT IS YOUR DIAGNOSIS?

WORKING DIAGNOSES
- Electromyography is often unhelpful even in the most overt cases of muscle disease
- The forearm ischemic exercise test excludes a disorder of glycogenolysis, including McArdle's, whereas the marked rise in plasma ammonia confirms the presence of myoadenylate deaminase
- The normal plasma urate suggests that there is no evidence of myogenic hyperuricemia (2)
- The absence of either raised acute-response proteins or viral titers makes an inflammatory or viral cause unlikely

FURTHER INVESTIGATIONS
- Muscle biopsy: Normal fiber types and distribution. A hint of increased aggregations of subsarcolemmal mitochondria noted
- The Sudan Black stain showed a small but significantly increased amount of fat in the type 1 fibers
- Electron microscopy confirmed this (see figure) and showed that the mitochondria contained dense inclusions of unknown origin
- Muscle mitochondrial respiratory chain complexes and carnitine palmitoyl transferase II (CPTII) activities; results expressed as a ratio to citrate synthase to correct for mitochondrial enrichment:

Complex I	0.135 (0.104–0.268)
Complex II/III	0.147 (0.040–0.204)
Complex IV	0.028 (0.014–0.034)
CPTII	23 (20–80)

- Acyl-carnitine profile by tandem mass spectrometry raised $C_{18:1}$, $C_{16:1}$, and $C_{14:1}$ species
- Tritiated water release from 9,10-tritium-labeled myristate, palmitate, and oleate by cultured fibroblasts expressed as a percentage of control: myristate C_{14} 76%, palmitate C_{16} 70%, oleate C_{18} 22% (kindly performed by Dr. S. Olpin, Sheffield Children's Hospital, UK)
- Muscle acyl coenzyme A (CoA) dehydrogenase specific activities (nmol/min/mg protein)

Octanoyl CoA	4.7 (1.2–7.0)
Palmitoyl CoA	1.7 (3.4–12.0)
Oleoyl CoA	0.9 (2.6–9.2)

FIGURE   ELECTRON MICROGRAPH TO SHOW PATIENT'S MUSCLE ACCUMULATION OF NEUTRAL LIPID DROPLETS (courtesy of Professor D. Landon, Institute of Neurology)

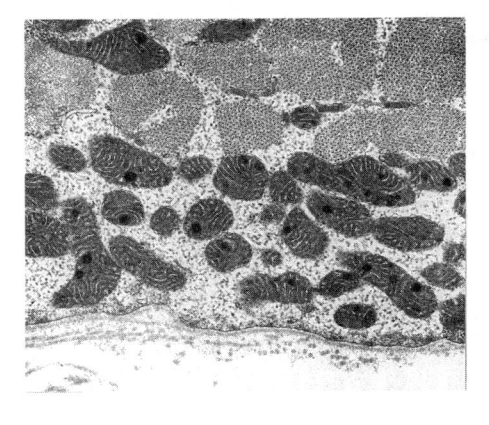

FINAL DIAGNOSIS: Very-long-chain acyl-CoA dehydrogenase (CAD) deficiency

DISCUSSION
- The muscle biopsy suggests a mitochondrial problem
- Analysis of the activity of the mitochondrial respiratory complexes, however, was normal
- Although small, the accumulation of fat within the muscle fibers suggested an impaired ability to utilize long-chain fatty acids
- This was supported by the finding of an abnormal long-chain acyl-carnitine profile in the patient's blood and by the impaired ability of fibroblasts to metabolize lipid, particularly the very-long-chain fatty acid oleate
- Reduced VLCAD activity was confirmed by direct enzymatic assay using oleate (C_{18}) as substrate
- The loss of some activity using palmitate (C_{16}) is explained by the overlapping substrate specificity of VLCAD and medium-chain acyl-CoA dehydrogenase (MCAD)
- Subsequently, the patient was shown to be heterozygous for a G520A mutation in exon 7 and to be heterozygous for a 2–base pair deletion in exon 4 at cDNA position 249–250 T of the gene for VLCA (kindly performed by Dr. B. Andresen, University of Aarhus, Denmark)

POINTS TO REMEMBER
- VLCAD has only recently been recognized. The enzyme was discovered and purified in 1992, allowing the first definitive diagnosis to be made utilizing immunoprecipitation and blotting techniques in 1993. In retrospect, it has been shown that previous cases of long-chain acyl-dehydrogenase deficiency described by Hale and others in the 1980s were in fact cases of VLCAD
- VLCAD deficiency may present in childhood as hypoketonemic hypoglycemia with cardiac involvement or later in life, as with CPTII, with isolated skeletal muscle involvement, recurrent rhabdomyolysis, and myoglobinuria associated with fasting and/or exercise. As with CPTII deficiency, muscle biopsy may be normal or have only minimal fat accumulation if the patient is not fasted and is well
- Acyl-carnitine profiles invariably suggest the diagnosis with $C_{14:1}$ acyl-carnitine being raised. Hypocarnitinemia is common
- Unlike CPTII or other disorders of fatty acid oxidation (such as MCAD and long-chain 3-hydroxyacyl-CoA dehydrogenase), no prevalent pathogenic mutation is known, though the most frequently observed mutations appear to disrupt folding and assembly of VLCAD, thereby disrupting the necessary spacial and vectorial organization of VLCAD with CPT II and the trifunctional protein
- Patients with VLCAD deficiency are well advised to avoid periods of fasting (>12 h), and a high-protein drink before bed is useful to sustain gluconeogenesis. A low-fat diet is prudent with occasional assessment of the fat-soluble vitamins. If at any time lipid needs to be given this should be in the form of medium-chain triglyceride. At times of intercurrent infection, maintenance of adequate carbohydrate intake either orally or intravenously is imperative to maintain blood glucose concentrations

REFERENCES
1. Clark JB, Bates TE, Boakye P, et al. Investigation of mitochondrial defects in brain and skeletal muscle. In Turner AJ, Batchelard H, eds. A practical approach to the investigation of metabolic disease. Oxford, UK: IRL Press at Oxford University Press, 1996:151–74.
2. Mineo IM, Kono N, Hara N, et al. Myogenic hyperuricemia. N Engl J Med 1987;317:75–80.
3. Munnich A, Rustin P, Rotig A, et al. Clinical aspects of mitochondrial disorders. J Inherit Metab Dis 1992;15:448–55.

4. Wanders RJA, Vreken P, Den Boer MEJ, et al. Disorders of mitochondrial fatty acid β-oxidation. J Inherit Metab Dis 1999;22: 442–87.
5. Andressen BS, Olpin S, Poorthuis BJ, et al. Clear correlation of genotype with disease phenotype in very-long-chain acyl-CoA dehydrogenase deficiency. Am J Hum Genet 1999; 64:479–94.
6. Land JM. Aspects of skeletal muscle biochemistry in health and disease. In: Karparti G, Hilton-Jones D, Griggs RC, eds. Disorders of voluntary muscle, 7th ed. Cambridge, UK: Cambridge University Press, in press.

THIS CASE WAS PRESENTED BY DR. JOHN M. LAND, DR. SIMON J.R. HEALES, AND DR. IAIN HARGREAVES, NEUROMETABOLIC UNIT, NATIONAL HOSPITAL FOR NEUROLOGY AND NEUROSURGERY, LONDON, UK

40. Drowsiness and confusion in an 83-y-old woman

PRESENTATION

HISTORY
- An 83-y-old woman was admitted with a 1-d history of drowsiness and confusion
- She had a long history of dysphagia due to a benign esophageal stricture
- She had tolerated solids until 1 wk ago
- For the past day she had not even been able to swallow liquids

PAST MEDICAL HISTORY
- Carcinoma of the bladder 35 y ago: cystectomy and ureterosigmoidostomy
- Hysterectomy 15 y ago
- Amputation of a toe 7 y ago
- Resection of a gangrenous bowel 5 y ago
- Dilation of benign esophageal stricture 5 y ago

FAMILY HISTORY
- No significant family history known

SOCIAL HISTORY
- Difficult to obtain because of confusion, but it is known that she lives alone

ON EXAMINATION
- Hyperventilating
- Very dehydrated
- Abdomen scarred like a battlefield
- She was admitted under the care of the surgeons

WHAT ARE YOUR PROVISIONAL DIAGNOSES?

WHAT INVESTIGATIONS WOULD YOU REQUEST?

PROVISIONAL DIAGNOSES
- Worsening of esophageal stricture
- Diabetes mellitus (in view of dehydration, hyperventilation, and drowsiness) was considered, but excluded by a normal capillary glucose value for blood glucose obtained on a point-of-care instrument in the emergency room

INITIAL INVESTIGATIONS
Plasma electrolytes, glucose, osmolality, blood gases, and a full blood count were requested, together with an electrocardiogram and chest and abdominal radiographs

Plasma
Sodium	154 mEq/L	(154 mmol/L)
Potassium	3.0 mEq/L	(3.0 mmol/L)
Bicarbonate	6.0 mEq/L	(6.0 mmol/L)
Urea	53 mg/dL	(19 mmol/L)
Creatinine	1.2 mg/dL	(105 µmol/L)
Glucose	128 mg/dL	(7.1 mmol/L)
Osmolality	338 mOsm/kg	(338 mOsm/kg)

Blood
pH	7.1	(H^+ 80 nmol/L)
pCO_2	20 mmHg	(2.6 kPa)
pO_2	118 mmHg	(15.8 kPa)

PROGRESS
- An intravenous infusion was initiated with 3 L of dextrose saline, each liter containing 80 mEq (80 mmol) KCl, prescribed to be infused over 12 h
- A surgical resident noted the results of the investigations and considered that lactic acidosis may account for the presentation

---

WHAT IS YOUR DIAGNOSIS?

WHAT FURTHER INVESTIGATIONS WOULD YOU REQUEST?

PROGRESS AND FURTHER INVESTIGATIONS
- The laboratory was contacted to measure lactic acid because of the unexplained metabolic acidosis
- Latest results of investigations were reviewed, with the clinical team taking note of the patient's current clinical state (she was well perfused)
- Diabetes mellitus and severe renal failure could be reasonably excluded as a cause of metabolic acidosis, as could poisoning
- Drug therapy was also reviewed

WORKING DIAGNOSIS
- Partly compensated metabolic acidosis of undefined type and unknown cause

FURTHER INVESTIGATIONS
The laboratory initiated measurement of chloride and calculation of anion gap:

Plasma
Chloride	133 mEq/L	(133 mmol/L)
Anion gap	18 mEq/L	(18 mmol/L)

The anion gap was calculated by:

$$\text{Anion gap} = (Na^+ + K^+) - (Cl^- + HCO_3^-)$$

Either conventional or SI units can be used in this formula.

- The cause of the acidosis was hyperchloremic metabolic acidosis (with normal anion gap)
- Plasma lactate was measured subsequently in a stored sample and found to be normal

FINAL DIAGNOSIS
- Review of past medical history revealed that a long-standing ureterosigmoidostomy had been performed for carcinoma of bladder. The patient had been prescribed oral bicarbonate to combat the acidosis that follows this procedure. The bicarbonate could not be swallowed because of the stricture and had not been reinstated after admission
- The final diagnosis of the acidosis was therefore hyperchloremic (normal anion gap) metabolic acidosis due to transplantation of the ureters into the colon

FURTHER PROGRESS
- The patient was treated with 500 mL of 1.26% $NaHCO_3$ (with added KCl) over 2 h
- This brought her pH to 7.22 (H^+ 60 nmol/L) during the day, and another infusion of bicarbonate brought her pH to 7.40 (H^+ 40 nmol/L) by the following day
- Her plasma chloride fell to 116 mEq/L (116 mmol/L)
- She was stabilized to allow dilation of her stricture
- This was successful and she was re-established on oral bicarbonate treatment

PATHOPHYSIOLOGY
- Transplantation of the ureters into the colon results in urine being exposed to the absorptive processes of the large bowel
- Urinary chloride is reabsorbed in exchange for bicarbonate, leading to reduced plasma bicarbonate and increased plasma chloride

POINTS TO REMEMBER

- Lactic acidosis, although frequently suspected as a cause of unexplained metabolic acidosis, is not often confirmed, especially in well-perfused patients
- In elucidating the cause of metabolic acidosis, careful consideration must be given to the patient's:
  - Clinical condition (including past medical history)
  - Drug therapy
- Metabolic acidoses are often considered difficult to diagnose. Most are easily diagnosed if attention is given particularly to diabetes mellitus, renal failure, and poisoning
- Measurement of chloride and anion gap are important in the diagnosis of unexplained metabolic acidosis
- Causes of hyperchloremic acidosis include:
  - Diarrhea
  - Ureterosigmoidostomy
  - Renal tubular acidosis
  - Treatment with carbonic anhydrase inhibitors
  - Recovery phase of diabetes mellitus
  - Aldosterone deficiency
  - Early uremic acidosis
  - Treatment with HCl, NH_4Cl, arginine HCl, or lysine HCl

REFERENCES

1. Adrougué HJ, Wilson H, Boyd AE, Suki WN, Eknoyan G. Plasma acid-base patterns in diabetic ketoacidosis. N Engl J Med 1982;307:1603–10.
2. Hooper J. Hyperchloraemic acidosis during recovery from hyperglycaemic diabetic emergencies. J R Soc Med 1996;89:600.
3. Walmsley RN, White GH. A guide to diagnostic clinical chemistry, 3rd ed. Oxford: Blackwell Scientific Publications, 1994.

THIS CASE WAS PRESENTED BY DR. JAMES HOOPER, THE ROYAL BROMPTON HOSPITAL, LONDON, UK

41. Sugar, water—and more

PRESENTATION

HISTORY OF PRESENTING COMPLAINT
- A 53-y-old woman with a bipolar disorder and mental retardation was admitted to the hospital after a fall
- An arthroplasty for a fractured neck of femur was performed and 24 h later she was transferred to the psychiatric team
- She was managed conservatively but 3 wk later collapsed

PAST MEDICAL HISTORY
- Three years previously the patient had been referred to a hospital physician because of hirsutism, night incontinence, and increasing hypernatremia and hyperosmolality
- Before attending this outpatient appointment, she was admitted to the hospital with confusion and dehydration
- She was described as a compulsive water drinker by her caregivers and had had her regular intake of up to 9 L of water per day restricted prior to admission

DRUG HISTORY
- She was receiving procyclidine, clopixol, carbamazepine, norethisterone, and lithium before her first admission

ON EXAMINATION
- She was severely dehydrated and hypotensive
- She responded only to painful stimuli
- She was tachycardic and cyanotic

INITIAL INVESTIGATIONS

	AT TIME OF COLLAPSE		PREVIOUS ADMISSION	
Sodium	190 mEq/L	(190 mmol/L)	164 mEq/L	(164 mmol/L)
Potassium	3.6 mEq/L	(3.6 mmol/L)	3.6 mEq/L	(3.6 mmol/L)
Urea	87 mg/dL	(31.2 mmol/L)	64 mg/dL	(22.7 mmol/L)
Creatinine	2.5 mg/dL	(225 µmol/L)	2.9 mg/dL	(255 µmol/L)
Osmolality	405 mOsm/kg	(405 mOsm/kg)	337 mOsm/kg	(337 mOsm/kg)
Glucose	297 mg/dL	(16.5 mmol/L)	279 mg/dL	(15.5 mmol/L)
Lithium	Not measured		2.24 mEq/L	(2.24 mmol/L)
Urine osmolality	Not measured		154 mOsm/kg	(154 mOsm/kg)

WHAT ARE YOUR PROVISIONAL DIAGNOSES?

WHAT FURTHER INVESTIGATIONS ARE INDICATED OR SHOULD HAVE BEEN CARRIED OUT?

PROVISIONAL DIAGNOSES
- Dehydration
- Diabetes mellitus
- Lithium toxicity
- Primary polydipsia and diabetes insipidus (DI)

FURTHER INVESTIGATIONS

Ketones	Not detected	
Calcium	9.5 mg/dL	(2.37 mmol/L)
Albumin	3.2 g/dL	(32 g/L)
Urine microbiology	*Escherichia coli* detected	

Additional test results that may have provided useful information at the time of collapse include arterial blood hydrogen ion (pH), blood gases, bicarbonate, and lactate

Polyuria and polydipsia were documented on her previous admission when a water deprivation test had been carried out

	PLASMA OSMOLALITY (mOsm/kg)	URINE OSMOLALITY (mOsm/kg)
Basal sample	335	154
After 8-h dehydration	336	168
After vasopressin (DDAVP)	334	217

HAVE YOUR MADE ANY CHANGES TO YOUR PROVISIONAL DIAGNOSIS?

WHAT IS YOUR FINAL DIAGNOSIS?

FINAL DIAGNOSIS: Lithium-induced nephrogenic diabetes insipidus associated with diabetes mellitus and septicemia

MANAGEMENT
- She was rehydrated with repeated infusions of 5% dextrose and given a short-acting insulin to stabilize her blood glucose concentration
- Four weeks later she was euvolemic and her serum sodium, creatinine, and glucose concentrations had returned to normal
- She was discharged into the community for 24-h care

COMMENTARY
- In this patient the diagnosis of diabetes insipidus (DI) was overlooked at the time of her acute surgical admission and subsequent referral for psychiatric care. It was exacerbated by her mental deterioration and inability to respond to thirst, which resulted in gross dehydration. Her clinical condition was complicated by poor control of her coexisting diabetes mellitus and the development of an infection leading to septicemia. At the time of collapse, it was thought that her hyperosmolar coma was due to nonketotic hyperglycemia
- Patients with this condition usually develop severe hyperglycemia [blood glucose concentrations >900 mg/dL (>50 mmol/L)] with extreme dehydration and a high plasma osmolality but no ketosis. This patient fulfilled these criteria with the exception of a very high blood glucose concentration
- Nephrogenic DI was diagnosed 3 y earlier on the basis of her response to water deprivation over a period of 8 h. Although the basal plasma osmolality was elevated it did not increase during the test. Failure of the plasma osmolality to rise in response to water deprivation may indicate fluid intake during the test. This patient was observed carefully throughout the test and did not drink. Before the test her urine osmolality was consistently lower than the plasma osmolality and did not increase with either water deprivation or DDAVP. The diagnosis made was nephrogenic DI secondary to lithium therapy

KEY FEATURES
The clinical features of DI are:
- Thirst
- Polyuria
- Nocturia

Causes of polyuria include:
- Diabetes mellitus
- Renal disease
- Hypercalcemia
- Hypokalemia
- Excessive inappropriate thirst and drinking
    Primary polydipsia
- Inadequate secretion of osmoregulated vasopressin
    Central-cranial DI
- Impaired renal response to vasopressin
    Nephrogenic DI

BIOCHEMISTRY
- Vasopressin or antidiuretic hormone (ADH) is synthesized in the hypothalamus and released into the circulation from the posterior pituitary gland. Vasopressin regulates water excretion through its action on the renal tubule and in healthy individuals acts in conjunction with the sensation of thirst to control plasma osmolality within a very narrow range
- The primary defect in DI is the inability to concentrate urine due to either impaired vasopressin release (central DI) or to an impaired renal response (nephrogenic DI)

DIAGNOSIS OF DIABETES INSIPIDUS
- Document urine volume on at least three occasions to confirm polyuria (defined as a urine volume >3 L per day)
- Exclude other causes of polyuria such as diabetes mellitus, renal failure, hypercalcemia, and hypokalemia by measuring serum glucose, creatinine, calcium, potassium, and osmolality
- Measure urine osmolality

In patients with DI, the urine osmolality is decreased and less than the plasma osmolality, which is usually accompanied by hypernatremia

A random urine osmolality >750 mOsm/kg excludes the diagnosis

- Water deprivation test
    This test is used to confirm the presence of DI and to differentiate between central and nephrogenic forms

A urine osmolality of less than 300 mOsm/kg after 8 h dehydration confirms the diagnosis of DI. A subsequent increase in urine osmolality after DDAVP to a value of at least 750 mOsm/kg is consistent with central DI, whereas patients with nephrogenic DI do not respond

LITHIUM-INDUCED NEPHROGENIC DI
- Nephrogenic DI is a common complication of lithium therapy
- Symptomatic polyuria occurs in 20–40% of patients receiving lithium, and DI affects up to 12%
- Lithium has been reported to cause down-regulation of the water channel protein of the renal collecting duct, aquaporin, as well as reduction of cyclic adenosine monophosphate (cAMP) in the distal tubule and collecting duct and the impairment of cAMP effect on the tubule
- The risks of prescribing lithium must therefore be balanced against the benefits of its use in controlling affective disorders

POINTS TO REMEMBER
- Be aware of disorders that can present with either polyuria or polydipsia or both
- Understand the differences between central DI and nephrogenic DI
- Recognize the side effects of lithium therapy including nephrogenic DI

REFERENCES

1. Adam P. Evaluation and management of diabetes insipidus. Am Fam Physician 1997;55:2146–52.
2. Baylis PH. Investigation of suspected hypothalamic diabetes insipidus. Clin Endocrinol 1995;43:507–10.
3. McGregor DA, Baker AM, Appel RG, Ober KP, Zaloga GP. Hyperosmolar coma due to lithium-induced diabetes insipidus. Lancet 1995;346:413–7.
4. Letters in reply to Reference 3. Lancet 1995;346:1428–9.
5. McKenna K, Thompson C. Osmoregulation in clinical disorders of thirst appreciation. Clin Endocrinol 1998;49:139–52.

THIS CASE WAS PRESENTED BY MRS. RUTH LAPWORTH, WILLIAM HARVEY HOSPITAL, ASHFORD, KENT, UK

42. Muscle pain and hypolipidemic therapy

PRESENTATION

PRESENTING COMPLAINT
- A 48-y-old man attended the lipid clinic for regular review and complained of long-standing angina pains and breathlessness
- He also complained of generalized limb muscle aches that had been present for several years
- The patient volunteered that movement made the pain in his limbs worse and that this made sleep difficult

PAST MEDICAL HISTORY
- The patient had suffered an acute myocardial infarction (AMI) at the age of 37 y
- His lipid results at that time were consistent with a diagnosis of heterozygous familial hypercholesterolemia (see table)
- He was under review by the cardiothoracic surgeons with a view to heart transplantation
- Since the AMI, he had been on lipid-lowering therapy—usually dual therapy with a 3-hydroxy-3-methylglutaryl coenzyme A (HMG CoA) reductase inhibitor and a fibric acid derivative. He had also significantly reduced his intake of dairy products

SOCIAL HISTORY
- The patient is the product of a consanguineous marriage and is a first-generation immigrant to the United Kingdom from East Asia
- The patient is a Muslim

MEDICATION
- Simvastatin, fenofibrate, captopril, furosemide with amiloride, cimetidine, tramadol, acetaminophen, warfarin, bisacodyl, and isosorbide mononitrate

FAMILY HISTORY
- The patient's mother had died prematurely from coronary heart disease
- His sister, two cousins, and two sons had biochemical evidence of heterozygous familial hypercholesterolemia. One of these cousins had developed premature coronary heart disease

ON EXAMINATION
- Overweight: height 5 ft 9 in (1.75 m), weight 183 pounds (83 kg), body mass index 27.1 kg/m^2
- Bilateral arcus senilis and bilateral small achilles tendon xanthoma. No xanthelasma
- Slow, shuffling gait—limited by pain and difficulty in rising from a sitting position
- Muscles in all four limbs and the chest wall were tender on palpation
- No limitation to passive limb movement

WHAT DO YOU SUSPECT IS THE CAUSE OF THE MUSCLE PAIN?

HOW WOULD YOU CONFIRM THIS?

PROVISONAL DIAGNOSES
- The clinical details given indicate that the patient probably has a generalized myositis. The most likely cause at this stage is his medication—the patient is on two drugs known to cause myositis. However, other important causes of myopathy cannot be excluded

INITIAL INVESTIGATIONS
Fasting values obtained in clinic
Plasma

Total cholesterol	317 mg/dL	(8.2 mmol/L)
Triglyceride	150 mg/dL	(1.7 mmol/L)
High-density lipoprotein cholesterol	33.6 mg/dL	(0.87 mmol/L)
Low-density lipoprotein cholesterol	255 mg/dL	(6.6 mmol/L)
Sodium	137 mEq/L	(137 mmol/L)
Potassium	4.1 mEq/L	(4.1 mmol/L)
Urea	23 mg/dL	(8.1 mmol/L)
Urate	8.1 mg/dL	(0.48 mmol/L)
Creatinine	1.3 mg/dL	(115 mmol/L)
Glucose	103 mg/dL	(5.7 mmol/L)
Total Protein	7.6 g/dL	(76 g/L)
Albumin	3.8 g/dL	(38 g/L)
Total bilirubin	0.4 mg/dL	(7 µmol/L)
Alanine aminotransferase	41 IU/L	(41 IU/L)
Alkaline phosphatase	768 IU/L	(768 IU/L)
γ-Glutamyltransferase	64 IU/L	(64 IU/L)
Creatine kinase (CK)	948 IU/L	(948 IU/L)
Free thyroxine	1.4 ng/dL	(18.3 pmol/L)
Thyroid-stimulating hormone	2.3 µU/mL	(2.3 mU/L)

FURTHER INFORMATION
There had been several occasions over the previous few years when elevated plasma CK had necessitated stopping hypolipidemic therapy. It had been noted on these occasions that stopping drug therapy had little effect on his symptoms and resulted in only a very slow fall in plasma enzyme activity
- From the above results and information available, hypothyroidism and alcohol can be excluded as a cause of myopathy
- The alkaline phosphatase appears elevated relative to the γ-glutamyltransferase value—indicating the possibility that increased amounts of bone isoenzyme are contributing to the total activity

---

WHAT OTHER INVESTIGATIONS WOULD YOU PERFORM TO CONFIRM THE LIKELY SOURCE OF THE CREATINE KINASE?

WHAT OTHER TESTS WILL HELP YOU FURTHER INVESTIGATE THE CAUSE OF THE MUSCLE PAIN?

MUSCLE PAIN AND HYPOLIPIDEMIC THERAPY

CK isoenzymes
- CK isoenzyme analysis will help confirm the source of the elevated plasma activity
- In this case, isoenzyme analysis was carried out using an electrophoretic method:
  - All of the elevated plasma concentration could be accounted for by a rise in the CK-3 (CK-MM) fraction
  - This result indicates that skeletal muscle is the likely source of the raised enzyme activity [as opposed to myocardium, which will result in elevated CK-2 (CK-MB) or smooth muscle—CK-1 (CK-BB)]

Osteomalacia is a common (but often forgotten) cause of myopathy, and an assessment of bone mineral metabolism was performed:

Total calcium	6.76 mg/dL	(1.69 mmol/L)
Albumin	4.1 g/dL	(41 g/L)
Corrected calcium	6.7 mg/dL	(1.67 mmol/L)
Inorganic phosphate	2.5 mg/dL	(0.8 mmol/L)
Magnesium	2.2 mg/dL	(0.91 mmol/L)
Alkaline phosphatase	768 IU/L	(768 IU/L)
Alkaline phosphatase isoenzymes (electrophoresis)	Predominantly bone	
25-Hydroxycholecalciferol	<5 ng/mL	(<12.5 nmol/L)
Parathyroid hormone	499 pg/mL	(52.2 pmol/L)

- These biochemical findings are consistent with a diagnosis of osteomalacia
- The alkaline phosphatase isoenzyme result confirms bone as the likely source of the raised plasma enzyme activity
- The hyperparathyroidism (secondary) is appropriate for the patient's hypocalcemia
- Four years previously his total calcium, corrected calcium, phosphate, and alkaline phosphatase values were all reported as normal

OTHER INVESTIGATIONS
- X-ray examination of the pelvis failed to reveal the presence of Looser's zones

TREATMENT
- The patient received one intramuscular injection of calciferol (300,000 units) followed by oral supplementation with 800 U of calciferol and 1000 mg of calcium daily

PROGRESS

	SEPTEMBER 23	OCTOBER 13	MARCH 14	SEPTEMBER 5
Adjusted calcium [mg/dL (mmol/L)]	6.7 (1.67)	7.9 (1.97)	9.7 (2.42)	9.7 (2.42)
Alkaline phosphatase (IU/L)	768	646	216	197
CK (IU/L)	638	–	856	1046

- His alkaline phosphatase fell from a peak of 768 IU/L to 197 IU/L and his plasma calcium

(adjusted for albumin concentration) rose from 6.7 mg/dL (1.67 mmol/L) to 9.7 mg/dL (2.42 mmol/L) over the next year
- As can be seen from the table, there has been no response of plasma CK activity to adequate treatment of the osteomalacia
- This lack of response suggests that osteomalacic myopathy may not have been the sole cause of the elevated CK and that it was a drug-related myositis

This patient's risk factors for osteomalacia include:
- Low-dairy-product diet (as part of the lifestyle changes adopted after AMI)
- Pigmented skin
- Wearing clothing that allows only minimal skin exposure to ultraviolet light
- Little exercise out of doors—because of chest pain and breathlessness, he was unable to walk far and spent most of his day indoors

DIAGNOSIS
- Osteomalacia—probably multifactorial
- Drug-induced myositis—hypolipidemic agents
- Heterozygous familial hypercholesterolemia

KEY FEATURES
PRESENTATION
- Myopathy and muscle weakness are important clinical features of osteomalacia, which may be seen in up to 30% of patients presenting with other biochemical evidence of the problem
- Muscle pain is a recognized side effect of therapy with HMG CoA reductase inhibitors and fibric acid derivatives

BIOCHEMISTRY
- A significant proportion of the daily requirement for vitamin D is supplied through production of precursors in the skin by the action of ultraviolet light on 7-dehydrocholesterol
- When consumption of dairy products is low, intake of both calcium and vitamin D will also be reduced. This reduction leaves the individual even more reliant on production of vitamin D precursors through skin irradiation with ultraviolet light
- In this case, skin irradiation was severely limited—partly by the type of clothing worn by the patient and partly through his severe incapacity, which meant that he spent very little time out of doors
- A further risk factor in those with severely compromised cardiac function is renal failure. This results in reduced conversion of 25-hydroxycholecalciferol to 1,25-dihydroxycholecalciferol

DIAGNOSIS
- Consider osteomalacia as a possible cause of myopathy—even if other causes of muscle symptoms may also be present
- Measure calcium, phosphate, alkaline phosphatase, and 25-hydroxycholecalciferol
- Measurement of γ-glutamyltransferase may be unhelpful in distinguishing between bone and liver sources of alkaline phosphatase in patients on multidrug therapy
- Analysis of alkaline phosphatase isoenzymes may be helpful if there is any doubt as to the likely source of a raised plasma enzyme activity. In some individuals, measurement of 1,25-dihydroxycholecalciferol and parathyroid hormone may be required to properly characterize calcium status

TREATMENT
- Supplementation with calciferol (1-alfacalcidol if vitamin D deficiency is secondary to renal disease)
- Calcium supplementation if plasma calcium is considered sufficiently low

POINTS TO REMEMBER
- Muscle pain (myositis) is an uncommon but important side effect of treatment with HMG CoA reductase inhibitors and fibric acid derivatives
- Muscle pain and weakness is a common (but often forgotten) clinical feature of osteomalacia
- The incidence of this drug-induced myositis increases when HMG CoA reductase inhibitors and fibric acid derivatives (or other cytochrome P450 3A4 inhibitors) are used together
- Although the most obvious cause of the muscle pain and elevated CK activity in this patient was a drug-induced myositis, it is important to bear in mind other possible causes in individuals with other multiple risk factors

REFERENCES
1. Glerup H, Mikkelsen, Poulsen L, Hass E, Overbeck S, Andersen H. Hypovitaminosis D myopathy without biochemical signs of osteomalacic bone involvement. Calcif Tissue Int 2000;66:419–24.
2. Goulding A. Lightening the fracture load: growing evidence suggests many older New Zealanders would benefit from more vitamin D [Editorial]. N Z Med J 1999;112:329.
3. Gruer PJ, Vega JM, Mercuri MF, Dobrinska MR, Tobert JA. Concomitant use of cytochrome P450 3A4 inhibitors and simvastatin. Am J Cardiol 1999;84:811–5.
4. Hilton-Jones D. Metabolic and endocrine myopathies. In: Weatherall DJ, Ledingham JGG, Warrell DA, eds. Oxford textbook of medicine. Oxford: Oxford University Press, 1996:4168–9.
5. Reginato AJ, Falasca GF, Pappu R, McKnight B, Agha A. Musculoskeletal manifestations of osteomalacia: report of 26 cases and literature review. Semin Arthritis Rheum 1999;28:287–304.

THIS CASE WAS PRESENTED BY DR. GEOFFREY SMITH, SOUTHERN COMMUNITY LABORATORIES LTD, CHRISTCHURCH, NEW ZEALAND

43. A young psychotic long-distance traveler

PRESENTATION

HISTORY OF PRESENTING COMPLAINT
- A 28-y-old man who had traveled from South America complained of feeling unwell during the 30-mile (48-km) trip home from the international airport
- He became increasingly agitated and aggressive; the taxi driver gladly deposited him at the nearest hospital
- While waiting in the accident and emergency department, he collapsed with seizures

PREVIOUS MEDICAL HISTORY
- Due to the patient's presentation no history could be obtained

ON EXAMINATION
- Due to the very aggressive behavior of the patient and the occurrence of grand mal seizures, examination was perfunctory and episodic
- He smelled of alcohol
- He had psychomotor agitation
- He was disorientated in place and time and he was delusional
- His blood pressure was 180/120 mmHg and pulse 110
- Temperature 39.8 °C
- No electrocardiogram (ECG) or X-ray was able to be taken
- He was incoherent and violently uncooperative, but complained of chest pain

---

WHAT IS YOUR PROVISIONAL DIAGNOSIS?

WHAT INVESTIGATIONS WOULD YOU REQUEST?

PROVISIONAL DIAGNOSIS
- Delirium tremens
- Delusional mental illness
- Epilepsy
- Drug-induced psychosis

INITIAL INVESTIGATIONS

Arterial gases

pH	7.17	(H^+ 68 nmol/L)
pCO_2	28.6 mmHg	(3.8 kPa)
pO_2	93.2 mmHg	(12.4 kPa)
Bicarbonate	13 mEq/L	(13 mmol/L)

Serum

Sodium	140 mEq/L	(140 mmol/L)
Potassium	5.6 mEq/L	(5.6 mmol/L)
Urea	16 mg/dL	(5.7 mmol/L)
Creatinine	1.1 mg/dL	(95 µmol/L)
Heroin or morphine	Not detected	
Amphetamines	Not detected	
Cocaine	2.76 mg/dL	(27.6 mg/L)
Ethanol	210 mg/dL	(4.6 mmol/L)

WHAT FURTHER INVESTIGATIONS, IF ANY, WOULD YOU REQUEST?

HAVE YOU CHANGED YOUR PROVISIONAL DIAGNOSIS?

WHAT IS YOUR DIAGNOSIS?

FURTHER INVESTIGATIONS
- Determine whether cocaethylene is present
    - Cocaethylene 1.5 mg/L
- X-ray to determine whether subject is a body-packer
- ECG to determine effect of cocaine on cardiac function
- Troponin I or T to exclude myocardial damage or infarct (subsequently found to be normal)

FINAL DIAGNOSIS: Drug-induced psychosis secondary to leakage of contents when body-packing cocaine

OUTCOME
- The patient continued to be extremely violent and nearly unapproachable
- He had a cardiac arrest during a further set of seizures and was pronounced dead
- At autopsy he was found to have ~120 packages of powder with a latex covering in his gastrointestinal tract; three of these had ruptured

KEY FEATURES
PRESENTATION
- Recreational cocaine use may involve the following routes of administration: nasal "snorting," smoking, injection, or ingestion. Crack (cocaine base) is typically smoked and cocaine powder is snorted
- Recreational drugs vary in purity. The intravenous route requires a tenth of the dose ingested for similar potency
- Body-stuffers and body-packers nonrecreationally ingest cocaine
- Body-stuffers ingest non- or poorly packaged drug to avoid detection. Due to bioavailability they are at risk of significant symptoms
- Body-packers ("mules") are drug smugglers who ingest packaged drug in a bid to avoid detection. Packages are the size of a very large grape; all are prone to leakage; some types are worse than others. Complete release of one package would result in a lethal dose
- Cocaine causes cardiac arrhythmia, ischemia through coronary artery vasospasm, and cardiac arrest
- Seizures are associated with higher doses
- A typical ECG pattern may demonstrate a prolonged QRS complex and QTc duration
- A plain X-ray may not detect packages depending on coating; latex coating may be seen as white densities

BIOCHEMISTRY
- Severe acid-base disturbances are associated with marked cocaine toxicity
- Hypoventilation is secondary to trauma or reduced mental status
- Metabolic acidosis is related to underperfusion
- Acidemia can exacerbate cocaine cardiac toxicity, for example, conductance
- Cocaine in the presence of ethanol is hepatically transesterified to the active metabolite cocaethylene (ethyl ester of benzoylecgonine)
- Cocaethylene has a longer half-life than cocaine and is approximately equipotent
- Cocaine is metabolized to benzoylecgonine, ecgonine methyl ester, and several minor metabolites, including anhydroecgonine if smoked
- Low plasma cholinesterase activity may predispose a person to increased cocaine toxicity due to diminished clearance (minor pathway)

DIAGNOSIS
- The history of long-distance travel and the onset of symptoms during or after flight, typically in a young person, should raise suspicion of body-packing
- Physical examination for signs and symptoms
- Check for acidemia
- Typically, a urine screen for drugs of abuse would enable detection of: cocaine, heroin, amphetamine, and cannabis mules
- ECG to monitor cardiac function and detect typical changes
- Plain X-ray to determine number of packages. A negative X-ray does not preclude body-packing
- Ethanol may be a complicating factor

TREATMENT
- All suspected body-packers should be admitted
- If hyperthermic (>41 °C) actively cool with ice
- Control convulsions with diazepam
- Control hypertension; avoid β-blockers
- Nitroglycerine for chest pain; thrombolytics if required
- Bicarbonate for broad complex ECG changes also helps address any acidemia (present in ~30% of cases)
- Further bicarbonate as required to correct acidosis
- Whole bowel irrigation for gut decontamination with polyethylene glycol orally or by nasogastric tube; continue until rectal effluent is clear and there are no more packages. Monitor urea and electrolytes
- Asymptomatic patients can be given activated charcoal
- Symptomatic patients should have operative intervention. Endoscopic removal is not advisable due to the risk of package damage
- Danger of bowel obstruction
- Observe post-treatment to ensure passage of two package-free stools before discharge
- Other abused substances may affect cocaine pharmacodynamics and pharmacokinetics

POINTS TO REMEMBER
- Unusual behavior in a long-distance air passenger could be because of drugs
- The index of suspicion is higher if they are under 35 years of age
- Body packing is not uncommon
- Leakage from packages will release fatal amounts of drug
- The best screen for detection or exclusion is a rapid urine test
- Active intervention may require patient sedation and must not exacerbate the situation

REFERENCES
1. Bailey DN. Effects of drugs and cocaine metabolites on cocaine and cocaethylene binding to human serum in vitro. Ther Drug Monit 1997;19:427–30.
2. Brookoff D, Rotondo MF, Shaw LM, Campbell EA, Fields L. Cocaethylene levels in patients who test positive for cocaine. Ann Emerg Med 1996;27:316–20.
3. Bogusz MJ, Althoff H, Erkens M, Maier RD, Hofmann R. Internally concealed cocaine: analytical and diagnostic aspects. J Forensic Sci 1995;40:811–5.
4. Cone EJ, Tsadik A, Oyler J, Darwin WD. Cocaine metabolism and urinary excretion after different routes of administration. Ther Drug Monit 1998;20:556–60.

5. Hoffman RS, Henry GC, Howland RS, et al. Association between life-threatening cocaine toxicity and plasma cholinesterase. Ann Emerg Med 1992;21:247–53.
6. Hoffman RS, Smilkstein MJ, Goldfrank LR. Whole bowel irrigation and the cocaine body packer. Am J Emerg Med 1990;8:523–7.
7. Hollander JE. Cocaine associated myocardial infarction. J R Soc Med 1996;89:443–7.
8. Jatlow P, McCance EF, Bradberry CW, Elsworth JD, Taylor JR, Roth RH. Alcohol plus cocaine: the whole is more than the sum of its parts. Ther Drug Monit 1996;18:460–4.
9. June R, Aks SE, Keys N, Wahl M. Medical outcome of cocaine bodystuffers. J Emerg Med 2000;18:221–4.
10. Karch SN, Stephens B, Ho CH. Relating cocaine blood concentrations to toxicity—an autopsy study of 99 cases. J Forensic Sci 1998;43:41–5.
11. McCance EF, Price LH, Kosten TR, Jatlow PI. Cocaethylene: pharmacology, physiology and behavioural effects in humans. J Pharmacol Exp Ther 1995;274:215–23.
12. McCarron MM, Wood JD. The cocaine "body-packer" syndrome. JAMA 1983;250:1417–20.
13. Perez-Reyes M, Jeffcoat AR, Myers M, et al. Comparison in humans of the potency and pharmacokinetics of intravenously injected cocaethylene and cocaine. Psychopharmcology (Berl) 1994;16:428–32.
14. Singhal PC, Rubin RB, Peters A, Santiago A, Neugarten J. Rhabdomyolysis and acute renal failure associated with cocaine abuse. J Toxicol Clin Toxicol 1990;28:321–30.
15. Stevens DC, Campbell JP, Carter JE, Watson WA. Acid-base abnormalities associated with cocaine toxicity in emergency department patients. J Toxicol Clin Toxicol 1994;32:31–9.
16. Wang RY. pH-dependent cocaine-induced cardio-toxicity. Am J Emerg Med 1999;17:364–9.

THIS CASE WAS PRESENTED BY DR. IAN WATSON, UNIVERSITY HOSPITAL, AINTREE, LIVERPOOL, UK

44. Convulsions in a neonate

PRESENTATION
- Male infant of consanguineous issue
- Normal delivery and weight
- Clinical examination unremarkable except for micrognathia
- Floppy

ON EXAMINATION
- Clinically jaundiced with bilirubin of 5.6 mg/dL (102 µmol/L), which was predominantly conjugated
- Ketonuric using test stick
- Convulsions that were unresponsive to phenytoin

BIOCHEMICAL INVESTIGATION

pH	7.30	(H^+ 50 nmol/L)
pCO_2	11 mmHg	(1.5 kPa)
Bicarbonate	7 mEq/L	(7 mmol/L)
Glucose	32 mg/dL	(1.8 mmol/L)

ON EXAMINATION AT 3 MONTHS OF AGE
- Unconscious with occasional fits

WHAT IS YOUR PROVISIONAL DIAGNOSIS?

WHAT INVESTIGATIONS WOULD YOU REQUEST?

PROVISIONAL DIAGNOSES

- Metabolic acidosis possibly due to:
  - Liver disease
  - Pump failure
  - Hypoxia
  - Amino acidopathy
  - Organic acidemia
  - Congenital lactate acidosis
- This baby presented with hypoglycemia that can be associated with prematurity or a history of maternal diabetes. Similarly, the elevated bilirubin could be consistent with immaturity of uridine 5'-diphosphate glucuronyl transferase. The baby was not premature, nor did the mother have diabetes
- Hypoglycemia and hyperbilirubinemia with a metabolic acidosis in a baby with severe neurological irritation as evidenced by severe convulsions suggests an inborn error of metabolism

FURTHER INVESTIGATIONS

Lactate	97 mg/dL	(10.8 mmol/L)
Ammonia	308 µg/dL	(220 µmol/L)

- These findings in addition to the reduced plasma glucose could be associated with severe end-stage liver disease, possibly secondary to an inborn error of metabolism, but this was inconsistent with the clinical examination. Assessment of cardiac function was unremarkable, and pCO_2 and hemoglobin saturation were normal, eliminating pump failure and hypoxia from the differential diagnosis
- Neonatal plasma ammonia concentrations are higher than in children. A plasma ammonia of 308 µg/dL is significantly elevated and when associated with an elevated lactate is consistent with an organic acidemia. In addition, urine thin-layer chromatography showed a generalized aminoaciduria

The results of the urine organic and amino acids were as follows:

	mg/24 h	µmol/24 h	REFERENCE RANGES [mg/24 h (µmol/24 h)]
Alanine	64	(712)	[5.6–14.3 (64–160)]
Lysine	272	(1867)	[1.5–20.4 (10–140)]
Threonine	100	(845)	[3.1–8.8 (26–74)]
Glycine	53	(700)	[10.7–42.4 (140–560)]
Proline	40	(347)	[<11.5 (<100)]
Arginine	21	(120)	[5.3 (30)]
Citrulline	83	(479)	[<20 (<10)]
Methylmalonic acid	Not detected		

WHAT FURTHER INVESTIGATIONS, IF ANY, WOULD YOU REQUEST?

HAVE YOU MADE ANY CHANGES TO YOUR PROVISIONAL DIAGNOSIS?

WHAT IS YOUR DIAGNOSIS?

WORKING DIAGNOSIS
- Congenital lactic acidosis

Using a sterile technique, fibroblasts were obtained from a full-skin-thickness punch biopsy. The fibroblast cells were grown in culture, and the activity of enzymes involved in congenital lactic acidosis measured

	PATIENT	NORMAL CONTROL
Pyruvate dehydrogenase (activated)	0.366[a]	[0.174]
Phosphoenolpyruvate carboxykinase	1.3[a]	[0.76]
Pyruvate carboxylase	0.004[a]	[0.195]

[a] Units of enzymatic activity: μmol of product/mg of protein/min

FINAL DIAGNOSIS : Congenital lactic acidosis due to a pyruvate carboxlyase (PC) deficiency (EC 6.4.1.1, McKusick 26615)

KEY FEATURES
- The near-normal urine glycine excretion tends to exclude biotin-responsive organic acidurias. Although several amino acids were increased, the concentration of citrulline was significantly elevated. The organic acids that are associated with hypoglycemia and a lactic acidosis were normal, including methylmalonic acid, elevated concentrations of which are associated with one of the commonest organic acidurias
- PC is a biotinylated mitochondrial enzyme that catalyzes the first committed step of gluconeogenesis, the conversion of pyruvate to oxaloacetate. Children with PC deficiency have lactic acidosis, hypoglycemia, and mental retardation
- The congenital PC deficiencies can be subdivided into at least two types. In the first and less-severe form, the urine amino acid profile is normal, and PC activity is detectable and cross-reacts with antibody against PC. This variant is called the cross-reacting material (CRM+). In the second form of PC deficiency, death occurs within the first 3–6 mo of life, there is no residual PC activity, and there is no cross-reactivity with anti-PC; hence, this form is termed CRM– (1)
- In the CRM– form, urine citrulline concentrations are markedly increased due to a secondary defect in ureagenesis. Low concentrations of mitochondrial oxaloacetate give rise to decreased concentrations of aspartate, which reduce urea production from ammonia, in turn causing the increase in urine citrulline and blood ammonia concentrations (2)
- There is marked heterogeneity in PC deficiency. In one patient who was still living at 8 y old, fibroblast culture showed 2–5% of the normal PC activity. This child had brief episodes of metabolic acidosis and was moderately mentally retarded, yet continued to develop language and motor skills (3). This case is atypical

MANAGEMENT
- Treatment of PC deficiency is aimed at increasing mitochondrial oxaloacetate by administration of citrate and aspartate. In one patient, this treatment caused a fall in plasma lactate and ketones with a normalization in all plasma amino acids except arginine; cerebrospinal fluid amino acids remained abnormal. Treatment improved metabolic control, but there was no clinical improvement (4)
- Early diagnosis of PC deficiency is important in that it may be part of the multiple carboxylase deficiency, some cases of which are responsive to biotin supplementation. Potentially if diagnosed

early and of the CRM+ variety, some of these patients will, if given biotin supplementation, show a clinical and biochemical improvement

POINTS TO REMEMBER
- In very ill children, it is important to take specimens of urine and blood for further investigations
- The investigations of metabolic disorders should be carried out while the patient is on a normal diet
- Other causes of lactic acidosis in a neonate are cardiac failure and hypoxia. The association of lactic acidosis and elevated serum ammonia in a child with neurological irritation suggests a metabolic defect
- Metabolic defects require a thorough, ordered, and enthusiastic investigation because achieving a diagnosis may assist in the treatment of other siblings

REFERENCES
1. Robinson BH, MacKay N, Chun K, Ling M. Disorders of pyruvate carboxylase and the pyruvate dehydrogenase complex. J Inherit Metab Dis 1996;19:452–6.
2. Greter J, Gustafsson J, Holme E. Pyruvate-carboxylase deficiency with urea cycle impairment. Acta Paediatr Scand 1985;74:982–6.
3. Stern HJ, Nayar R, Depalma L, Rifai N. Prolonged survival in pyruvate carboxylase deficiency: lack of correlation with enzyme activity in cultured fibroblasts. Clin Biochem 1995;28:85–9.
4. Ahmad A, Kahler SG, Kishnani PS, et al. Treatment of pyruvate carboxylase deficiency with high doses of citrate and aspartate. Am J Med Genet 1999;87:331–8.

THIS CASE WAS PRESENTED BY MR. PETER TIMMS, ST BARTHOLOMEW'S HOSPITAL, LONDON, UK

45. Found in a coma

PRESENTATION

HISTORY OF THE PRESENTING COMPLAINT
- A 77-y-old man was brought to the emergency department by his daughter
- He had been found by her in a coma in his apartment

DIRECT QUESTIONING (OF DAUGHTER)
- Episodes of hypoglycemic attack when not eating regularly
- Eating habit variable since death of his wife in 1987
- Episodes normally occurred at night when he woke up with nocturia (six times a night)
- No polyuria during the day
- Hypoglycemic attacks always preceded by tingling of lips, palpitation, and double vision
- According to his daughter, during the attacks he is incoherent
- He is a heavy drinker

PAST MEDICAL HISTORY
- Appendectomy in 1944
- Hernia repair in October 2000
- Abdominal X-ray and computed tomography (CT) scan in 1994 for the investigation of hypoglycemia showed no abnormality of pancreas

FAMILY HISTORY
- History of diabetes mellitus in the family
- Father died of a myocardial infarction at 70 y
- Mother fit and well at age 99 y

DRUG HISTORY
- Nothing of note

ON EXAMINATION
- Pale and sweaty

IMMEDIATE INVESTIGATIONS AND TREATMENT
- Finger-prick test for glucose was 19.8 mg/dL (1.1 mmol/L)
- He recovered rapidly with administration of glucose

WHAT IS YOUR PROVISIONAL DIAGNOSIS?

WHAT INVESTIGATIONS WOULD YOU REQUEST?

PROVISIONAL DIAGNOSES

Hypoglycemia due to:
- Diabetes or diabetic treatment or abuse, for example, insulin or oral hypoglycemic agents
- Alcohol abuse
- Endocrine disorder, such as Addison's disease or hypopituitarism
- Insulinoma or pancreatic tumor [multiple endocrine neoplasia type 1 (MEN type 1)]
- Malignancy

INITIAL INVESTIGATIONS

Serum
Sodium	134 mEq/L	(134 mmol/L)
Potassium	4.1 mEq/L	(4.1 mmol/L)
Chloride	101 mEq/L	(101 mmol/L)
Bicarbonate	24 mEq/L	(24 mmol/L)
Calcium	9.8 mg/dL	(2.46 mmol/L)
Phosphate	3.2 mg/dL	(1.05 mmol/L)
Alkaline phosphatase	76 IU/L	(76 IU/L)
Alanine aminotransferase	31 IU/L	(31 IU/L)
Aspartate aminotransferase	34 IU/L	(34 IU/L)
γ-Glutamyltransferase	184 IU/L	(184 IU/L)
Glucose	36 mg/dL	(2.0 mmol/L)
Insulin	25 μU/mL	(150 pmol/L)
Cortisol at 0900 h	9.1 μg/dL	(250 nmol/L)
Growth hormone	Undetectable	

Urine
Ketones	Negative	

WHAT FURTHER INVESTIGATIONS, IF ANY, WOULD YOU REQUEST?

HAVE YOU MADE ANY CHANGES TO YOUR PROVISIONAL DIAGNOSIS?

WHAT IS YOUR DIAGNOSIS?

WORKING DIAGNOSES

Hyperinsulinemia due to:
- Insulinoma
- Insulin abuse
- Sulfonylurea (oral hypoglycemic agent) abuse
- MEN type 1

FURTHER INVESTIGATIONS

- Fasting glucose, insulin, and C-peptide measurement
- Sulfonylurea measurement
- Intestinal peptides
- Parathyroid hormone (PTH) and fasting calcium
- Abdominal CT scans

Serum
Glucose	45 mg/dL	(2.5 mmol/L)
Insulin	35 µU/mL	(210 pmol/L)
C-peptide	4.5 ng/mL	(1.5 nmol/L)
Sulfonylurea	<0.2 ng/mL	(< 0.2 µg/L)
Calcium	10.0 mg/dL	(2.50 mmol/L)
PTH	45 ng/mL	(4.7 pmol/L)

- Intestinal peptide screen: All peptides were within the reference range
- CT scan: A 1.4-cm region of increased vascular uptake was noted in the midbody of the pancreas
- A percutaneous transhepatic portal vein catheterization with venous sampling was performed to localize the tumor

FINAL DIAGNOSIS: Insulinoma

MANAGEMENT

- Short-term therapeutic trial of octreotide and diazoxide to give symptomatic relief
- Surgical removal of insulin-producing tumor

KEY FEATURES

PRESENTATION

- The complexity of the glucose regulatory mechanism provides a heterogeneity of causes for hypoglycemia syndrome
- Hypoglycemia is confirmed when plasma glucose concentration falls below 40 mg/dL (2.2 mmol/L)
- Insulinoma is a rare (incidence of 1 in 1,000,000 population per year) but important cause of hypoglycemia. Fasting hypoglycemia and its associated neuroglycopenia are the presenting symptoms of insulinoma
- Symptoms frequently occur in the morning before breakfast and in-between meal times. Hypoglycemia also occurs during or after rigorous exercise or dieting
- Inappropriate hyperinsulinemia is not demonstrated on all occasions
- Because 30–40% of the pancreatic tumors in MEN type 1 patients are insulinoma, the possibility of MEN must be explored in the initial presentation of hypoglycemia with hyperinsulinemia
- Sulfonylureas produce hypoglycemia with increase in insulin and C-peptide secretion. Differential diagnosis of insulinoma must exclude use of sulfonylurea as a cause of hypoglycemia

BIOCHEMISTRY
- Insulin is normally the only blood glucose–lowering hormone to circulate in significant amounts
- Insulin reduces hepatic glucose output and increases the uptake of glucose by the tissue
- It also suppresses the fatty acid release from the adipocytes and depresses the ketone production and thus lowers both glucose and ketone concentrations
- Because glucagon, adrenaline, growth hormone, and cortisol are insulin antagonists, deficiency of any of these can give rise to hypoglycemia
- Several factors determine the plasma insulin concentration. The most important is the arterial blood glucose concentration. Hypoglycemia from any other cause will inhibit endogenous production of insulin and C-peptide, and peripheral venous plasma concentration of insulin and C-peptide will fall to <5 mU/mL (<30 pmol/L) and <1 ng/mL (<0.3 nmol/L), respectively

DIAGNOSIS
- Depends on the demonstration of symptomatic hypoglycemia in the presence of inappropriately high plasma insulin and C-peptide. This combination is characteristic of endogenous hyperinsulinemia due to insulinoma or sulfonylurea poisoning
- Factitious hypoglycemia with hyperinsulinemia has low C-peptide concentration. Measurement of fasting insulin, C-peptide, and sulfonylurea will normally confirm the diagnosis
- In many instances, the hypoglycemia is intermittent and manifests only during prolonged fasting. This is true in patients with early mild inappropriate insulin secretion due to islet cell adenoma. In such instances, prolonged fasting is helpful to confirm the diagnosis of hypoglycemia
- During prolonged fasting in patients with insulinoma, Whipple's triad (low plasma glucose, symptoms of neuroglycopenia, and amelioration of symptoms by glucose administration) is demonstrated in 95–98% of cases within 72 h of fast
- Measurement of PTH, calcium, and pituitary and gut peptides will help to confirm insulinoma associated with MEN type 1

TREATMENT
- All patients with insulinoma require octreotide or diazoxide cover to give symptomatic relief
- Surgical removal of the tumor is the ultimate option

POINTS TO REMEMBER
- Hypoglycemia due to hyperinsulinemia is associated with a low concentration of ketones. Thus, measurement of ketones should be performed early in the investigation of hypoglycemia
- Because incidence of intentional or accidental use of oral hypoglycemic agent and insulin is common, sulfonylurea abuse is an important differential diagnosis
- Insulin values should be interpreted together with the glucose concentration at the time of sampling

REFERENCES
1. Marks V. Hypoglycaemia and insulinomas. In: Besser GM, Thorner OM, eds. Clinical endocrinology, 2nd ed. London: Times Mirror International, 1994:20.1–20.16.
2. Cryer PE, Polonsky KS. Glucose homeostasis and hypoglycaemia. In: Wilson JD, Foster DW, Kronenberg HM, Larsen PR, eds. Williams textbook of endocrinology, 9th ed. Philadelphia: WB Saunders, 1998:939–71.
3. Gagle RF. Multiple endocrine neoplasia. In: Wilson JD, Foster DW, Kronenberg HM, Larsen PR, eds. Williams textbook of endocrinology, 9th ed. Philadelphia: WB Saunders, 1998:1627–49.

THIS CASE WAS PRESENTED BY MRS. KATE NOONAN, THE ROYAL LONDON HOSPITAL, BARTS AND THE LONDON NHS TRUST, LONDON, UK

46. Metabolic encephalopathy

PRESENTATION

HISTORY OF PRESENT COMPLAINT
- A 32-y-old woman was admitted to the hospital
- She had become confused over a period of 24 h
- Before that, she had had a headache and felt unwell

PAST MEDICAL HISTORY
- She had suffered from occasional generalized tonic-clonic seizures since childhood
- She also experienced episodic severe throbbing headaches attributed to common migraine
- She was otherwise healthy, mentally normal, and without any dietary fads
- She had five seizures in the 12 mo preceding the decision to start her on sodium valproate at 200 mg twice a day, increasing to 400 mg twice a day after 5 d

DRUG HISTORY
- Her only medication was the oral contraceptive Brevinor (norethindrone with ethinyl estradiol)

FAMILY HISTORY
- There was no known relevant family history

It is now 1500 h on Sunday. You ask the chemical pathologist for advice

---

WHAT ADVICE SHOULD BE GIVEN?

WHAT INVESTIGATIONS WOULD YOU REQUEST?

WHAT DIAGNOSIS WOULD YOU CONSIDER?

COURSE AFTER ADMISSION
- She was given a single dose of dexamethasone and mannitol
- The brain computed-tomography scan was normal, and the patient's condition was attributed to a post-ictal state
- She became less drowsy, but remained confused
- Thirty-six hours after admission she experienced a series of focal seizures and became profoundly unconscious, with extensor posturing of all limbs

INITIAL INVESTIGATIONS
Serum

Bilirubin	1.0 mg/dL	(17 µmol/L)
Aspartate aminotransferase	55 IU/L	(55 IU/L)
Alkaline phosphatase	90 IU/L	(90 IU/L)
Albumin	4.0 g/dL	(40 g/L)
γ-Glutamyltransferase	50 IU/L	(50 IU/L)
Prothrombin time	12.0 s	
Activated partial thromboplastin time	40 s	

- By chance, the neurology senior house officer was on a rotation that included pediatrics and recalled a lecture on inborn errors of urea synthesis
- The plasma ammonia concentration (measured as nitrogen) was found to be 1172 µg/dL (650 µmol/L)

WHAT DOES THIS SUGGEST?

DOES THIS EXCLUDE A HEPATIC OR METABOLIC CAUSE FOR THE CLINICAL PICTURE?

WHAT FURTHER INVESTIGATIONS SHOULD BE PERFORMED?

- Encephalopathy is often observed with this degree of hyperammonemia, although the correlation between symptoms and the plasma concentration among individuals and at different times is poor

FURTHER INVESTIGATIONS
- The same ammonia concentration was recorded on a second plasma specimen, which was rapidly transported and analyzed
- The urinary orotate-creatinine ratio was 863 mg/g creatinine (625 × 10^{-3} mmol/mol creatinine)
- The plasma and urinary organic acid chromatograms showed no diagnostic pattern
- The lack of accumulation of other amino acids in the urea cycle suggests a defect high in the pathway
- The raised orotate concentration in urine suggests deficiency of ornithine carbamoyltransferase (OCT), rather than carbamoylphosphate synthetase or an organic acidemia

FINAL DIAGNOSIS : These results suggest a deficiency of OCT (EC 2.1.3.3)

- Liver specimens obtained by percutaneous biopsy showed a low activity of 339 μmol/g wet wt/h (reference range 2200–10,700); the activity of carbamoylphosphate synthetase was 199 μmol/g wet wt/h (reference range 76–515)

SUBSEQUENT COURSE
- The patient was ventilated, but despite a fall of plasma ammonia (measured as nitrogen) to 112 μg/dL (80 μmol/L), she remained decerebrate and ventilation was discontinued after 5 d

MANAGEMENT
- How easy is it to measure plasma ammonia in your laboratory at 1500 h on Sunday?
- Do laboratories perpetuate the fallacy of liver function tests?
- Do we pay too much attention to unnecessary tests performed out of hours and not enough to necessary tests not performed?

KEY FEATURES
OCT DEFICIENCY AND VALPROATE THERAPY
- This women had partial OCT deficiency, and the terminal episode was precipitated by the administration of valproate
- The intermittent seizures before treatment was started may have been a manifestation of the primary metabolic disorder, but the wide range of severity of this condition (*1*) makes it impossible to distinguish it retrospectively from idiopathic epilepsy
- There are two reports of the interaction of valproate and OCT deficiency in children (*2,3*), and family histories suggestive of genetic defects occur in three other series describing valproate toxicity (*4–6*)
- Literature searches have not revealed similar reports in adults
- This interaction may be due to the effect of valproate on carbamoylphosphate synthesis (*7,8*), the step preceding OCT in urea synthesis

LIVER FUNCTION TESTS
- It is not clear whether descriptions of fatal liver failure after valproate therapy and those of symptomatic hyperammonemia refer to the same condition, and the appropriate metabolic studies should be added to conventional liver function tests and histological examination in the investigation of suspected valproate toxicity

POINTS TO REMEMBER
- Know the metabolic causes of encephalopathy, including hyperammonemia and hypoglycemia
- Be able to plan the further investigation after the identification of hyperammonemia
- Understand the interaction between the inborn error of metabolism and environmental factors including catabolism and valproate administration in the development of hyperammonemia
- Identify the information resources that will help in the investigation of rare diseases

REFERENCES
1. Walser M. Urea cycle disorders and other hereditary hyperammonaemic syndromes. In: Stanbury JB, Wyngaarden JB, Fredrickson DS, Goldstein JL, Brown MS, eds. The metabolic basis of inherited disease. New York: McGraw-Hill, 1983:402–8.
2. Tripp JH, Hargreaves T, Anthony PP, et al. Sodium valproate and ornithine carbamyltransferase deficiency. Lancet 1981;1:1165–6.
3. Hjelm M, de Silva LVK, Seakins JWT, Oberholzer VG, Rolles CJ. Evidence of inherited urea cycle defect in a case of fatal valproate toxicity. Br Med J 1986;292:23–4.
4. Ware S, Millward-Sadler GH. Acute liver disease associated with sodium valproate. Lancet 1980;2:1110–3.
5. Sucy FJ, Balistreri WF, Buchino JJ, et al. Acute hepatic failure associated with the use of sodium valproate. N Engl J Med 1979;300:962–6.
6. Scheffner D, König St. Valproate hepatotoxicity. Lancet 1987;1:389–90.
7. Hjelm M, Oberholzer VG, Seakins J, Thomas S, Kay JDS. Valproate-induced inhibition of urea synthesis and hyperammonaemia in healthy subjects. Lancet 1986;2:859.
8. Hjelm M, Oberholzer V, Seakins J, Thomas S, Kay JDS. Valproate inhibition of urea synthesis. Lancet 1987;2:923–4.

THIS CASE WAS PRESENTED BY DR. JONATHAN KAY, JOHN RADCLIFFE HOSPITALS, OXFORD, UK

47. A diabetic emergency

PRESENTATION

HISTORY OF PRESENTING COMPLAINT
- A 58-y-old Caucasian man with type 2 diabetes presented to the accident and emergency department with severe abdominal pain radiating to his back
- The patient stated that this was the worst pain he had ever experienced. The pain had worsened over the last 6 h. He felt nauseous, thirsty, and feverish and had vomited once

PAST MEDICAL HISTORY
- The patient was diagnosed with type 2 diabetes at 50 y of age; he had a strong family history, with both mother and father having type 2 diabetes
- He had been hypertensive for 15 y

DRUG HISTORY
- 2.5 mg bendrofluazide
- 100 mg atenolol
- 80 mg of gliclazide twice daily
- 500 mg of metformin daily (could not tolerate more)

SOCIAL HISTORY
- Bus driver
- Smoked 20 cigarettes a day, which he had done for years
- Drank 30 units of alcohol a week

ON EXAMINATION
- Alert but sweating, dry tongue
- Very tender over epigastrium with a rigid abdomen
- No gastrointestinal problems
- Electrocardiogram normal, but pulse 120/min

INITIAL INVESTIGATIONS
- Chest clear, no gas under diaphragm
- Body-mass index 30
- Point-of-care blood glucose 333 mg/dL (18.5 mmol/L)
- Urine ketones were negative

WHAT IS YOUR PROVISIONAL DIAGNOSIS?

WHAT INVESTIGATIONS WOULD YOU REQUEST?

PROVISIONAL DIAGNOSIS
- Metabolic complication of diabetes
- Possible myocardial infarction
- Pancreatitis
- Esophagitis or hiatus hernia
- Peptic ulcer

INVESTIGATIONS

Serum
Sodium[a]	145 mEq/L	(145 mmol/L)
Potassium[a]	4.0 mEq/L	(4.0 mmol/L)
Bicarbonate	18 mEq/L	(18 mmol/L)
Urea	12 mg/dL	(4.2 mmol/L)
Creatinine	1.2 mg/dL	(105 µmol/L)
Calcium	8.2 mg/dL	(2.05 mmol/L)
Albumin	3.9 g/dL	(39 g/L)
Bilirubin	0.7 mg/dL	(12 µmol/L)
Alkaline phosphatase	90 IU/L	(90 IU/L)
Alanine aminotransferase	50 IU/L	(50 IU/L)
Aspartate aminotransferase	70 IU/L	(70 IU/L)
γ-Glutamyltransferase	90 IU/L	(90 IU/L)
Amylase	1500 IU/L	(1500 IU/L)
Hemoglobin	10.7 g/dL	(107 g/L)
Leukocyte count	$16.0 \times 10^3/\mu L$	(16.0×10^9/L)

[a] Sodium and potassium were measured by direct ion-selective electrode (ISE) because sample was lipemic

FURTHER INVESTIGATIONS
Cholesterol	676 mg/dL	(17.5 mmol/L)
Triglyceride	6265 mg/dL	(70.8 mmol/L)

WHAT IS YOUR FINAL DIAGNOSIS?

FINAL DIAGNOSIS: Pancreatitis secondary to hypertriglyceridemia and alcohol abuse
- Hypertriglyceridemia enhanced or precipitated by an alcoholic binge
- Massive hypertriglyceridemia associated with frank pancreatitis is uncommon, but clinically important and under-recognized. It may arise as a result of severe genetic defects in lipolysis or, more commonly, from a moderate primary hypertriglyceridemia that is exacerbated by a secondary cause

TREATMENT
- The patient was given a low-calorie saline dextrose fluid replacement [**not** 5% dextrose-saline] with a sliding scale of insulin; initially 6 U/h
- The following day, the triglyceride was 1363 mg/dL (15.4 mmol/L), and cholesterol was 317 mg/dL (8.2 mmol/L). The electrolytes had normalized, and the amylase had fallen to 970 IU/L
- The patient was placed on regular insulin therapy at his follow-up hospital appointment

KEY FEATURES
- In the United States, most patients with acute pancreatitis have alcoholism or gallstones as an etiologic factor. Alcoholic pancreatitis develops in susceptible persons after heavy ethanol ingestion for many years. Chronic alcoholism may produce proteinaceous plugs in the small pancreatic ducts, causing atrophy of the acini drained by the obstructed duct. These chronic, irreversible pathologic changes antedate the first attack of acute pancreatitis, and 10% of alcoholics develop pancreatic insufficiency without a recognized acute attack
- A triglyceride concentration of 900–9000 mg/dL (10–100 mmol/L) may result from several causes. It may be secondary to alcohol abuse, diabetes, or other causes (see table); inherited causes include familial hypertriglyceridemia; the chylomicronemia syndrome with deficient lipoprotein lipase or apolipoprotein C-II activator; rarely, immunoglobulin binding to lipoprotein lipase, heparin, or C-II; and remnant (familial type III) hyperlipidemia
- Patients with untreated type 1 diabetes and untreated symptomatic type 2 diabetes have low adipose tissue or muscle lipoprotein lipase with a mild to moderate increase in triglyceride concentrations and decreased high-density lipoprotein (HDL) cholesterol concentrations. With insulin resistance and milder insulin deficiency, hypertriglyceridemia is caused by excess free fatty acids mobilized from adipose tissue that are re-esterified in the liver and secreted as endogenous very-low-density lipoproteins (VLDL). Activity of the enzyme lipoprotein lipase is low in people with untreated diabetes who have moderate to severe hypertriglyceridemia and impaired removal of VLDL from plasma
- Pancreatitis is a rare complication of triglyceride excess, generally occurring only when triglyceride concentrations increase beyond 2200 mg/dL (25 mmol/L), and major hypertriglyceridemia is a rare accompaniment to pancreatitis, with a prevalence of ~3% in each situation. But it is important because it resolves with conservative treatment, any underlying metabolic problem can be addressed later when all is stabilized, and inadvertent surgery for acute abdomen would be potentially disastrous because the combination of surgery and chemical pancreatitis (from free fatty acid release) will leave a legacy of contracting adhesions. The observant clinical biochemist will advise the surgical team of the unusual appearance of the admission blood sample, which, one hopes, may make surgeons think again and hold off laparotomy for the acute abdomen

- Pancreatitis may rumble on for months or years with several subclinical attacks, progressively reducing insulin availability and progressively moving type 2 diabetes towards type 1 diabetes
- The laboratory workers noted that the sample was lipemic (milky) and did not report the sodium [125 mEq/L (125 mmol/L)] or potassium [3.5 mEq/L (3.5 mmol/L)] measured routinely using the indirect ISEs on the analyzer. They reported electrolytes that were measured using a direct ISE. The low sodium measured on the analyzer gives rise to a condition known as pseudohyponatremia. Sodium is present in the plasma water compartment and not in the lipid compartment of a sample; when lipid is increased in a given sample volume, the water compartment (containing sodium) is reduced. This has the effect of lowering the measured sodium. This effect is also found with high protein concentrations, but this increase is not visually noticeable as with increased lipids. Pseudohyponatremia should be considered in all samples that produce an unexpectedly low sodium result
- This patient also had a low calcium [8.2 mg/dL (2.05 mmol/L)]. Hypocalcemia is found in more than half of patients presenting with acute pancreatitis. This hypocalcemia is transient, rarely falls below 6.8 mg/dL (1.7 mmol/L) or causes symptoms, and usually occurs within 36 h of the onset of pain. The cause of the hypocalcemia is poorly understood and various mechanisms have been proposed:
    - Damaged pancreatic cells → malabsorption of fats → calcium sequestration causing calcium soap deposits. Some studies conclude the amounts sequestered do not fully account for the observed falls in plasma calcium
    - Hypoalbuminemia as a result of shock
    - Glucagon release by damaged cells. Experimental glucagon infusion lowers plasma calcium by an unknown mechanism
    - Hypomagnesemia is commonly found in alcoholics because of poor diet
    - Release of unidentified substances that block the action of parathyroid hormone on bone resorption and/or inhibit parathyroid hormone secration
- The incidence of acute pancreatitis is reportedly higher in patients with primary hyperparathyroidism than in the general population, and it is suggested that long-standing hypercalcemia may be a minor cause of acute pancreatitis

TABLE
Some factors resulting in increased lipids

	MAIN LIPID FRACTION AFFECTED
Diabetes mellitus	↑ triglyceride, ↓ HDL, bias to small dense LDL
Alcohol abuse	↑ triglyceride
Chronic renal failure	↑ triglyceride and lipoprotein(a) [Lp(a)]
Drugs	↑ triglyceride and/or ↑ cholesterol
Nephrotic syndrome	↑ cholesterol and/or ↑ triglyceride
Hypothyroid	↑ cholesterol and/or LDL
Cholestasis	↑ cholesterol
Anorexia nervosa	↑ cholesterol

↑, increased; ↓, decreased; LDL, low-density lipoprotein

POINTS TO REMEMBER

- Hyponatremia (pseudo) may be present in a sample with a grossly elevated triglyceride concentration
- Hypertriglyceridemia is an important cause of pancreatitis
- Alcohol abuse is a common cause of pancreatitis
- A transient hypocalcemia often occurs in the acute stage of pancreatitis

REFERENCES

1. Miller JP. Serum triglyceride, the liver and the pancreas. Curr Opin Lipidol 2000;11:377–82.
2. Marshall JB. Acute pancreatitis: a review with emphasis on new developments. Arch Intern Med 1993;153:1185–98.
3. Steer ML. Classification and pathogenesis of pancreatitis. Surg Clin North Am 1989;69:467–80.
4. Wierzbicki AS, Reynolds TM. Familial hyperchylomicronaemia. Lancet 1996;348:1524–5.
5. Piolot A, Nadler F, Cavallero E, Coquard JL, Jacotot B. Prevention of recurrent acute pancreatitis in patients with severe hypertriglyceridemia: value of regular plasmapheresis. Pancreas 1996;13:96–9.
6. Chait A, Brunzell JD. Chylomicronemia syndrome. Adv Intern Med 1992;37:249–73.
7. Brunzell JD, Bierman EL. Chylomicronemia syndrome. Interaction of genetic and acquired hypertriglyceridemia. Med Clin North Am 1982;66:455–68.
8. Durrington PN. Secondary hyperlipidaemia. Br Med Bull 1990;46:1005–24.

THIS CASE WAS PRESENTED BY DR. GARRY JOHN, ST BARTHOLOMEW'S HOSPITAL, LONDON, UK

48. An uncommon urine dipstick reaction

PRESENTATION

HISTORY OF PRESENTING COMPLAINT
- A 48-y-old woman was seen by her general practitioner at home complaining of headache, diffuse abdominal pain, and vomiting
- Physical examination revealed no significant abdominal abnormality
- She was not jaundiced

PAST MEDICAL HISTORY
- Two months previously, she had undergone emergency admission to the hospital with a 2-d history of nausea, vomiting, and increasing confusion. Serum carbamazepine was 15.9 µg/mL (67 µmol/L). She recovered rapidly and was discharged on clobazam 10 mg twice a day, carbamazepine 300 mg twice a day, and vigabatrin 500 mg twice a day
- Three years previously she had developed focal epilepsy after removal of a parietal tumor. Her fits were only partially controlled by valproate and clobazam. Carbamazepine was started after 10 mo, with the dose being increased to 600 mg/d 2 mo later. About 1 y after her operation, she began to complain of new symptoms: headache, vomiting, and abdominal pain that were worse premenstrually and midcycle

SOCIAL AND FAMILY HISTORY
- Married university lecturer with two children
- No significant family history

INITIAL INVESTIGATIONS
- Dark brown urine. Urine dipstick test positive for urobilinogen; bilirubin negative
- Normal liver function tests. Normal hemoglobin and reticulocyte count

---

WHAT IS YOUR PROVISIONAL DIAGNOSIS?

WHAT INVESTIGATIONS WOULD YOU REQUEST?

PROVISIONAL DIAGNOSIS
- Urine dipstick tests for urobilinogen depend on its reaction with a diazonium salt to give a red color. Other substances, sometimes present in urine, may react similarly, of which the most important is porphobilinogen (PBG). Here, the absence of any evidence of liver disease or hemolysis, together with a history of an unexplained abdominal disorder that started when carbamazepine was added to her anticonvulsant treatment, raises the possibility that PBG is responsible for the positive dipstick test and requires exclusion of acute porphyria

FURTHER INVESTIGATIONS
Urine
    Porphobilinogen   194 mg/L   (849 μmol/L)
Feces
    Total porphyrin   85 μg/g dry wt   (140 nmol/g dry wt)
Erythrocytes
    PBG deaminase   30 nmol/h/mL of erythrocytes

WHAT IS YOUR FINAL PROVISIONAL DIAGNOSIS?

FINAL DIAGNOSIS: Acute intermittent porphyria (AIP)

MANAGEMENT
- Carbamazepine was withdrawn. Her abdominal symptoms and headache disappeared within a month and have not returned. One year later her urine PBG concentration had fallen to 1.2 mg/L
- Her family was screened for clinically latent AIP. Analysis of her genomic DNA identified a disease-specific mutation in exon 3 of her *HMBS* (hydroxymethylbilane synthase) gene, which encodes PBG deaminase. The mutation was present in one of her parents and in relatives descending from her maternal grandparents
- Affected relatives were given information about the acute porphyrias and measures that can be taken to diminish the risk of acute attacks

KEY FEATURES
- The diagnosis of AIP is established by the increased urinary PBG excretion with a normal fecal porphyrin concentration. No other investigations are required
- Her erythrocyte PBG deaminase activity was within the normal reference range, but within the overlap between that range and the range for AIP patients. This enzyme measurement is not useful as a front-line test for the diagnosis of an attack of acute porphyria and should be reserved for family studies
- Her clinical presentation was unusual in that she had recurrent abdominal symptoms with headache that caused discomfort rather than severe pain. Most patients with an attack of acute porphyria have very severe abdominal pain that requires opiate analgesia. Carbamazepine was clearly implicated as the provoking agent. This drug is known to be unsafe in the acute porphyrias
- About 20% of patients presenting with an attack of acute porphyria have no family history of the disorder, but almost all can be shown to have asymptomatic affected relatives in previous generations; the de novo mutation rate in AIP is low
- Urine dipstick tests for urobilinogen will detect greatly increased PBG concentrations, but are not sufficiently sensitive to be used to screen for increased urinary PBG

EPILEPSY AND ACUTE PORPHYRIA
- Convulsions occur in 5–20% of acute attacks of porphyria, often caused by hyponatremia
- The frequency of chronic epilepsy, in which fits are unrelated to acute attacks, is probably no greater in individuals with porphyria than in the general population
- Choice of anticonvulsants for the treatment of chronic epilepsy in individuals who have inherited one of the acute porphyrias is difficult because many are known to provoke acute attacks
- Anticonvulsants with a high risk of provoking an acute attack are phenobarbital and other barbiturates, phenytoin, primidone, ethosuximide, and carbamazepine. Benzodiazepines and probably valproate have a low risk and may be used with caution. Vigabatrin and gabapentin are also in the low-risk group, whereas lamotrigine, felbamate, and tiagabine are probably unsafe, but more clinical experience with these newer anticonvulsants is required
- Changes in anticonvulsant treatment may provoke acute attacks. PBG excretion should be monitored during the introduction of new drugs and their withdrawal considered if there is a consistent increase

POINTS TO REMEMBER
- Urine dipstick tests for urobilinogen are not specific; high concentrations of PBG also give positive reactions
- Acute porphyria should be considered as a cause of otherwise unexplained abdominal pain after the age of puberty, especially in any patient receiving anticonvulsants for epilepsy

REFERENCES
1. Anderson KE, Sassa S, Bishop DF, Desnick RJ. Disorders of heme biosynthesis: X-linked sideroblastic anemia and the porphyrias. In: Scriver CR, Beaudet AL, Sly WS, et al., eds. The molecular and metabolic bases of inherited disease, 8th ed. New York: McGraw-Hill, 2000:2961–3062.
2. Elder GH, Hift RJ, Meissner PN. The acute porphyrias. Lancet 1997;349:1613–7.
3. Gorchein A. Drug treatment in acute porphyria. Br J Clin Pharmacol 1997;44:427–34.
4. Bylesjo I, Forsgren L, Lithner F, Boman K. Epidemiology and clinical characteristics of seizures in patients with acute intermittent porphyria. Epilepsia 1996;37:230–5.
5. Hahn M, Gildemeister OS, Krauss GL, et al. Effects of new anticonvulsant medications on porphyrin synthesis in cultured liver cells. Neurology 1997;49:97–106.

THIS CASE WAS PRESENTED BY PROFESSOR GEORGE ELDER, UNIVERSITY OF WALES COLLEGE OF MEDICINE, CARDIFF, WALES, UK

Reference ranges for analytes in blood (adults)

- Values are for serum samples, unless otherwise stated
- Variations may exist between males and females and among people of different ages; only guideline ranges are given except in some specific cases
- There may be method-specific variations in reference ranges
- Readers are advised to consult reference texts and manufacturers' method data sheets to ascertain the nature of variations that should be noted

ANALYTE	CONVENTIONAL UNITS RANGE	UNITS	SI UNITS RANGE	UNITS
Acetone	0.3–2.0	mg/dL	0.05–0.34	mmol/L
Adrenocorticotrophic hormone (ACTH)				
0800 h	10–80	pg/mL	10–80	ng/L
2400 h	<10	pg/mL	<10	ng/L
Albumin	3.8–4.8	g/dL	38–48	g/L
Aldosterone				
Random sample	3.6–30.6	ng/dL	100–850	pmol/L
Overnight recumbent	3.6–16.2	ng/dL	100–450	pmol/L
Ambulant	3.6–30.6	ng/dL	100–850	pmol/L
Alanine aminotransferase	<40	IU/L	<40	IU/L
Alkaline phosphatase (total)	36–125	IU/L	36–125	IU/L
Alkaline phosphatase (bone isoform)	7–28	ng/mL	7–28	µg/L
Ammonia (as nitrogen)	15–45	µg/dL	11–32	µmol/L
Amylase	70–300	IU/L	70–300	IU/L
Androstenedione	52–343	ng/dL	1.8–12.0	nmol/L
Anion gap	12–20	mEq/L	12–20	mmol/L
α_1-Antitrypsin	80–180	mg/dL	0.8–1.8	g/L
Aspartate aminotransferase	<45	IU/L	<45	IU/L
Bicarbonate	21–30	mEq/L	21–30	mmol/L
Bilirubin (total)	<1.0	mg/dL	<17	µmol/L
Bilirubin (direct)	<0.3	mg/dL	<4	µmol/L
Calcitonin (male)	<100	pg/mL	<100	ng/L
Calcitonin (female)	<30	pg/mL	<30	ng/L
Calcium (total)	9.0–10.6	mg/dL	2.25–2.65	mmol/L
Calcium (corrected)	9.0–10.6	mg/dL	2.25–2.65	mmol/L
Calcium (ionized)	4.5–5.3	mg/dL	1.12–1.32	mmol/L
Carbon dioxide (pCO_2)	33–48	mmHg	4.39–6.38	kPa
Ceruloplasmin	20–45	mg/dL	0.20–0.45	g/L
Chloride	98–106	mEq/L	98–106	mmol/L

Analyte	Value	Unit	Value (SI)	Unit (SI)
Cholesterol total (ideal)	<200	mg/dL	<5.2	mmol/L
Cholesterol (high-density lipoprotein)	27–67	mg/dL	0.70–1.73	mmol/L
Cholesterol (low-density lipoprotein)	87–186	mg/dL	2.25–4.82	mmol/L
Copper	70–160	µg/L	11.0–25.1	µmol/L
Cortisol				
0800 h	4–28	µg/dL	138–772	nmol/L
2400 h	<50% of 0800 h value		<50% of 0800 h value	
C-peptide	0.78–1.89	ng/mL	0.26–0.62	nmol/L
Creatine kinase total activity	50–170	IU/L	50–170	IU/L
Creatine kinase MB isoenzyme (activity)	<16	IU/L	<16	IU/L
Creatine kinase MB isoenzyme (ratio)	<5	%	<5	%
Creatine kinase MB isoenzyme (mass)	<7	ng/mL	<7	µg/L
Creatinine	0.7–1.4	mg/dL	62–124	µmol/L
C-reactive protein (CRP)	<0.6	mg/dL	<6.0	mg/L
Cysteine (leukocytes)	<0.24	mg/g protein	<2.0	µmol/g protein
D-Dimer	<250	ng/mL	<250	µg/L
Dehydroepiandrosterone sulphate (DHAS)	81–370	µg/mL	2.2–10.0	µmol/L
1,25-Dihydroxycholecalciferol (1,25-diOH vit D)	15–60	pg/mL	15–60	ng/L
Estradiol				
Males	15–45	pg/mL	55–165	pmol/L
Females, early follicular	30–50	pg/mL	110–183	pmol/L
Females, midcycle	150–450	pg/mL	550–1650	pmol/L
Females, luteal	150–230	pg/mL	550–845	pmol/L
Females, postmenopausal	<55	pg/mL	<200	pmol/L
Ferritin (females)	6–110	ng/mL	6–110	µg/L
Ferritin (males)	20–260	ng/mL	20–260	µg/L
Folate (erythrocyte)	140–628	ng/mL	317–1422	nmol/L
Folate (serum)	3–16	ng/mL	7–36	nmol/L
Follicle-stimulating hormone (FSH)				
Males	1–9	mU/mL	1–9	U/L
Females, follicular	1–9	mU/mL	1–9	U/L
Females, midcycle	<50	mU/mL	<50	U/L
Females, luteal	1–8	mU/mL	1–8	U/L
Female, postmenopausal	>25	mU/mL	>25	U/L
Free androgen index (females)	<36.7		<7.5	

REFERENCE RANGES

Gastrin	10–85	pg/mL	10–85	ng/L
Glucose (fasting, plasma)	70–105	mg/dL	3.9–5.8	mmol/L
γ-Glutamyltransferase	5–75	IU/L	5–75	IU/L
Hemoglobin A$_{1C}$ (blood)	2.8–5.0	%	2.8–5.0	%
Hydrogen ion (pH) (arterial whole blood)	7.35–7.45		36–45	nmol/L
β-Hydroxybutyrate	0.22–0.88	mg/dL	21–85	μmol/L
25-Hydroxycholecalciferol (25-OH vit D)	13–50	ng/mL	33–125	nmol/L
17-Hydroxyprogesterone	67–400	ng/dL	2–12	nmol/L
Immunoglobulin A	65–375	mg/dL	0.65–3.75	g/L
Immunoglobulin E (total)	3–150	mIU/mL	3–150	IU/L
Immunoglobulin E (allergen specific)	<0.35	mIU/mL	<0.35	IU/L
Immunoglobulin G	800–1450	mg/dL	8.0–14.5	g/L
Immunoglobulin M	20–300	mg/dL	0.2–3.0	g/L
Immunoreactive trypsin (IRT)	<12	μg/dL	<120	μg/L
Insulin	6–24	μU/mL	42–167	pmol/L
Iron	33–140	μg/dL	6–25	μmol/L
Iron-binding capacity	250–400	μg/dL	44.8–71.6	μmol/L
Ketones	0.5–3.0	mg/dL	5–30	mg/L
Lactate	1.8–18.0	mg/dL	0.2–2.0	mmol/L
Lactate dehydrogenase	300–570	IU/L	300–570	IU/L
Lead (blood)	<40	μg/dL	<1.93	μmol/L
Luteinizing hormone (LH)				
Males	1–9	mU/mL	1–9	U/L
Females, follicular	1–10	mU/mL	1–10	U/L
Females, midcycle	<75	mU/mL	<75	U/L
Females, luteal	1–13	mU/mL	1–13	U/L
Females, postmenopausal	>16	mU/mL	>16	U/L
Magnesium	1.6–2.6	mg/dL	0.66–1.07	mmol/L
Myoglobin	<8	μg/dL	<80	μg/L
Osmolality	278–298	mOsm/kg	278–298	mOsm/kg
Oxygen (pO$_2$)	83–108	mmHg	11.1–14.4	kPa
Parathyroid hormone (intact)	10–65	pg/mL	1.1–6.8	pmol/L
Phosphate	2.5–4.7	mg/dL	0.8–1.5	mmol/L
Porphobilinogen deaminase (erythrocytes)			24–67	nmol/h/mL of erythrocytes

Potassium	3.5–5.1	mEq/L	3.5–5.1	mmol/L
Prolactin				
Males	3.0–14.7	ng/mL	3.0–14.7	µg/L
Females	3.8–23.2	ng/mL	3.8–23.2	µg/L
Renin				
Random sample	0.6–3.8	ng/h/mL	0.5–3.1	pmol/h/mL
Overnight recumbent	1.3–3.3	ng/h/mL	1.1–2.7	pmol/h/mL
Ambulant	3.4–5.5	ng/h/mL	2.8–4.5	pmol/h/mL
Rheumatoid factor	<20	mIU/mL	<20	IU/L
Sex hormone–binding globulin	1.42–4.85	mg/L	24–82	nmol/L
Sodium	135–145	mEql/L	135–145	mmol/L
α-Subunit	0.5–2.0	ng/mL	0.5–2.0	µg/L
Testosterone				
Males	280–1100	ng/dL	0.5–38.2	nmol/L
Females	<70	ng/dL	<2.5	nmol/L
Thyroid-stimulating hormone (TSH)	0.6–4.8	µU/mL	0.6–4.8	mU/L
Thyroxine (total)	4.7–11.2	µg/dL	60–145	nmol/L
Thyroxine (free)	0.74–2.05	ng/dL	9.6–26.5	pmol/L
Total protein	6.1–7.9	g/dL	61–79	g/L
Transferrin	200–360	mg/dL	2.0–3.6	g/L
Triglycerides	<185	mg/dL	<2.1	mmol/L
Triiodothyronine (total)	65–162	ng/dL	1.0–2.5	nmol/L
Triiodothyronine (free)	3.5–8.0	pg/mL	5.4–12.3	pmol/L
Troponin I (Access method)	<0.1	µg/L	<0.1	µg/L
Urate	2.5–7.1	mg/dL	0.15–0.42	mmol/L
Urea	7–18	mg/dL	2.5–6.5	mmol/L
Vitamin B_{12}	203–1300	pg/mL	150–960	pmol/L
Zinc protoporphyrin (erythrocytes)	<40	µg/dL	<0.7	µmol/L

Reference ranges for analytes in blood (infants)

- Values are for serum samples, unless otherwise stated
- Same criteria apply as for ranges in adults

ANALYTE	CONVENTIONAL UNITS RANGE	UNITS	SI UNITS RANGE	UNITS
Alkaline phosphatase	250–1000	IU/L	250–1000	IU/L
Aspartate aminotransferase	<80	IU/L	<80	IU/L
Creatinine	0.2–0.9	mg/dL	20–80	µmol/L
Ferritin	110–500	ng/mL	110–500	µg/L
Phosphate	4.0–6.4	mg/dL	1.3–2.1	mmol/L
Total protein	5.5–7.0	g/dL	55–70	g/L

Reference ranges for analytes in urine (adults)

- Variations may exist between males and females and among people of different ages; only guideline ranges are given
- There may be method-related variations in these ranges
- Readers are advised to consult reference texts and manufacturers' method data sheets to ascertain specific variations that should be noted

ANALYTE	CONVENTIONAL UNITS RANGE	UNITS	SI UNITS RANGE	UNITS
Adrenaline (99% confidence interval)	<40	µg/24 h	<220	nmol/24 h
Albumin	<80	mg/24 h	<80	mg/24 h
Amylase	<900	IU/24 h	<900	IU/24 h
Calcium	100–300	mg/24 h	2.5–7.5	mmol/24 h
Calcium-creatinine ratio	0–424	mg/g	0–1.2	mmol/mmol
Chloride	110–250	mEq/24 h	110–250	mmol/24 h
Copper	<50	µg/24 h	<0.79	µmol/24 h
Cortisol (free)	<109	µg/24 h	<300	nmol/24 h
Dopamine (99% confidence interval)	<569	µg/24 h	<3720	nmol/24 h
Homovanillic acid	1.4–8.8	mg/24 h	8–48	µmol/24 h
Hydrogen ion (pH)	4.5–8.0			
4-Hydroxy-3-methoxymandelic acid	1.4–6.5	mg/24 h	7–33	µmol/24 h
Noradrenaline (99% confidence interval)	<138	µg/24 h	<814	nmol/24 h
Orotate-creatinine ratio	0–5.5	mg/g	0–4	mmol/mol
Osmolality	50–1200	mOsm/kg	50–1200	mOsm/kg
Phosphate	495–1486	mg/24 h	16–48	mmol/24 h
Porphobilinogen	<2.4	mg/L	<10.6	µmol/L
Potassium	40–120	mEq/24 h	40–120	mmol/24 h
Protein-creatinine ratio	<175	mg/g	<20.0	mg/mmol
Sodium	100–250	mEq/24 h	100–250	mmol/24 h
Urea	12–20	g/24 h	430–710	mmol/24 h

Reference ranges for miscellaneous biochemical parameters

	CONVENTIONAL UNITS		SI UNITS	
	RANGE	UNITS	RANGE	UNITS
Cerebrospinal fluid glucose	40–70	mg/dL	2.2–3.9	mmol/L
Cerebrospinal fluid protein	8–32	mg/dL	80–320	mg/L
Fecal total porphyrin	<120	µg/g dry wt	<200	nmol/g dry wt

Reference ranges for therapeutic drug concentrations (serum concentrations)

	CONVENTIONAL UNITS		SI UNITS	
ANALYTE	RANGE	UNITS	RANGE	UNITS
Carbamazepine	4–10	µg/mL	17–42	µmol/L
Lithium	0.6–1.2	mEq/L	0.6–1.2	mmol/L
Phenytoin	5–20	µg/mL	40–79	µmol/L

Reference ranges for hematological parameters (adults)

- Variations may exist between males and females and among people of different ages; only guideline ranges are given
- There may be method-related variations in these ranges
- Readers are advised to consult reference texts and manufacturers' method data sheets to ascertain specific variations that should be noted

ANALYTE	CONVENTIONAL UNITS RANGE	UNITS	SI UNITS RANGE	UNITS
Activated partial thromboplastin time	25–35	s	25–35	s
Erythrocyte count	3.8–5.8	$\times 10^6/\mu L$	3.8–5.8	$\times 10^{12}/L$
Erythrocyte distribution width	11.5–14.5		11.5–14.5	
Erythrocyte sedimentation rate	1–15	mm/h	1–15	mm/h
Hemoglobin	11.5–16.5	g/dL	115–165	g/L
Leukocyte count	4.0–11.0	$\times 10^3/\mu L$	4.0–11.0	$\times 10^9/L$
Mean cell hemoglobin	27–35	pg	27–35	pg
Mean cell hemoglobin concentration	31–35	gHb/dL of erythrocytes	310–350	gHb/L of erythrocytes
Mean cell volume	76–96	μm^3	76–96	fL
Neutrophils	2.5–7.5	$\times 10^3/\mu L$	2.5–7.5	$\times 10^9/L$
Packed cell volume (hematocrit)	37–47	%	0.37–0.47	
Platelet count	150–450	$\times 10^3/\mu L$	150–450	$\times 10^9/L$
Platelet distribution width	14.9–16.9		14.9–16.9	
Prothrombin time	12.5–14.5	s	12.5–14.5	s
Reticulocyte count	24–84	$\times 10^3/\mu L$	24–84	$\times 10^9/L$

Cases by broad subject areas

Water and electrolyte	1, 2, 4, 12, 13, 14, 15, 16, 28, 33, 38, 40, 41
Acid-base	17, 18, 25, 33, 40, 43, 44
Liver	17, 21, 22, 26, 32, 42, 44, 46
Lipids and diabetes	10, 17, 25, 30, 32, 41, 42, 45, 47
Cardiac	2, 20
Calcium and bone	11, 24, 33, 35, 42
Thyroid	6, 7, 29, 31
Adrenal	1, 2, 3, 4, 5, 6, 10, 12, 13, 16, 19, 28, 41
Pituitary and fertility	5, 8, 9, 29
Screening and cancer	9, 10, 27, 29, 31, 35, 40, 45
Renal	13, 16, 25, 33, 35, 40, 41
Poisoning and therapeutic drugs	11, 18, 23, 26, 37, 38, 41, 43, 47
Pediatrics	2, 4, 5, 9, 21, 24
Hematology	7, 14, 22, 30
Gut and malabsorption	7, 15, 23, 29
Metabolic and miscellaneous	15, 17, 24, 25, 28, 34, 38, 39, 44, 46, 48

Index

Page numbers set in italics refer to figures or tables.

For measurements and terms that appear more than once in a particular case history, only the first occurrence is cited. The exception is diagnoses, which are cited on the page the Final Diagnosis appears, even if they occur earlier as provisional or working diagnoses.

A

Acardic twin, 131
Acetaminophen
    measurement, 79, 84, 124, 180
    nomogram for relating concentration to clinical outcome, *127*
    poisoning, 125
Acetone
    measurement, 181
    poisoning, 181
N-Acetyl-p-benzoquinonimine, 125
N-Acetylcysteine therapy, 125
Acidosis, 24, 76, 79, 84, 120, 171, 181, 193, 209, 215
Acute hemolytic anemia, 146
Acute intermittent porphyria, 232
Acyl fraction measurement (derived), 186
Addison's disease, 25
Adrenaline measurement, 46
Adrenocorticotropic hormone
    as cause of pigmentation, 16
    measurement, 12, 25, 90
    stimulation test, 21
Adrenocorticotropic hormone–releasing hormone stimulation test, 90
Alanine aminotransferase measurement, 52, 69, 83, 98, 104, 125, 130, 138, 144, 151, 170, 202, 218, 226
Albumin-creatinine ratio calculation, 61
Albumin measurement, 8, 24, 52, 66, 69, 104, 110, 125, 130, 138, 144, 151, 155, 159, 164, 170, 180, 196, 202, 222, 226
Alcohol
    excessive intake or abuse, 77, 125, 179, 217, 227
    measurement. *See* Ethanol measurement; Ethylene glycol measurement; Isopropanol measurement; Methanol measurement
Alcohol dehydrogenase inhibitor (fomepizole), 85
Alcoholic ketoacidosis
    as cause of acidosis, 79
    pathophysiology, *80*
Aldosterone
    antagonists, 5
    measurement, 3, 8, 20, 62, 74, 135, 171
Alkaline phosphatase
    isoenzyme determination, 203
    measurement, 52, 69, 83, 98, 104, 110, 113, 125, 130, 138, 144, 151, 159, 163, 170, 202, 218, 222, 226
Alkalosis, 76
Amino acids measurement, 98, 171, 214
Ammonia measurement, 186, 214, 222
Amphetamines screen, 208
Amylase measurement, 24, 78, 83, 110, 120, 138, 144, 226
Androgens, causes of excess in women, 37
Androstenedione measurement, 21, 36, 43
Anemia, 31, 146
Anion gap calculation, 79, 84, 180, 193
Antibodies to adrenal cortex measurement, 25
Antibodies to gastric parietal cells measurement, 30
Antibodies to intrinsic factor measurement, 30
Antibodies to thyroid peroxidase measurement, 25
Antinuclear factor measurement, 145
α_1-Antitrypsin measurement, 98, 151
Arthralgia, 103
Aspartate aminotransferase measurement, 70, 78, 95, 98, 110, 151, 164, 218, 222, 226
Aspartate therapy, 215
Autoimmune disease, 25, *27*, 31, 146, 149

Autoimmune hypothyroidism, 25, 149
Autoimmune polyglandular deficiency syndrome, 27
Autoimmune vitamin B$_{12}$ deficiency, 31

B

Bicarbonate
    measurement, 8, 24, 52, 62, 66, 73, 78, 103, 120, 124, 134, 159, 171, 180, 192, 208, 213, 218, 226
    therapy, 85, 193, 210
Biliary atresia, 99
Bilirubin measurement, 52, 69, 78, 83, 97, 104, 110, 125, 130, 138, 144, 151, 164, 202, 222, 226
Bladder carcinoma, 191
Blood film test, 30, 110
Body-packing, 209
Bone loss, 54
Brain natriuretic peptide, 59
Burkitt's lymphoma, 159

C

Calciferol therapy, 205
Calcitonin measurement, 47
Calcium
    24-h excretion measurement, 53
    measurement, 47, 52, 66, 103, 113, 121, 138, 140, 144, 155, 159, 163, 170, 180, 196, 203, 218, 226
    ratio to creatinine, 171
    therapy, 54, 111, 115, 140, 165, 171, 205
Carbamazepine measurement, 134, 231
Carnitine measurement, 186
Carnitine palmitoyl transferase II to citrate synthase ratio calculation, 187
Catecholamines
    measurement, 157
    metabolic pathway for production, *48*
Cerebral salt wasting, 59
Cerebrovascular accident, 1
Ceruloplasmin
    measurement, 151
    reduced synthesis, 153
CHARGE association, 116

Child (age 1–18 y) as subject of case history, 7, 19, 143, 159, 169
    See also Infant (<1 y) as subject of case history
Chloride measurement, 8, 24, 76, 79, 83, 171, 180, 193, 218
Cholestasis, 99
Cholesterol measurement
    high-density lipoprotein, 202
    low-density lipoprotein, 202
    total, 98, 103, 144, 202, 226
Chronic hepatitis C infection, 153
Chronic hyponatremia, 63
Citrate therapy, 215
Citrulline elevation, 215
Cocaethylene measurement, 209
Cocaine
    as cause of psychosis, 209
    body-packing, 209
    measurement, 208
Colon tumor, 163
Congenital adrenal hyperplasia
    as cause of accumulation of androgens, 37
    non salt losing, 21
    salt losing, 16
Congenital heart disease, 116
Congenital lactic acidosis, 215
Congestive heart failure, 9
Conn's syndrome, 3
Convulsions, 213
Copper
    excretion rate measurement, 152
    measurement, 152
Corticosteroid-binding globulin, 13
Corticotrophin-releasing hormone test, 13
Cortisol
    diurnal variation, 13
    measurement, 11, 20, 25, 36, 42, 62, 90, 135, 139, 218
C-peptide measurement, 219
C-reactive protein measurement, 24, 134, 151, 186
Creatine kinase
    isoenzyme CK-MB measurement, 93
    measurement, 24, 52, 71, 93, 121, 185, 202
Creatinine measurement, 3, 8, 20, 24, 52, 58, 61, 66, 69, 73, 78, 83, 95, 103, 110, 124, 130, 134, 138, 155, 159, 163, 175, 180, 192, 195, 202, 208, 226

INDEX

Cross-reacting material in pyruvate carboxylase deficiency, 215
Cushing's syndrome
    caused by adrenocorticotropic hormone overproduction with subsequent nodular adrenal hyperplasia, 91
    caused by glucocorticoid adenoma, 13
    in pregnancy, 13
    with nodular adrenal hyperplasia, 91
Cysteamine therapy, 171
Cysteine measurement, 171
Cystine toxicity, 172
Cystinosis
    as cause of constipation and failure to thrive, 171
    long-term complications, *172*

D

D-Dimer measurement, 151
Deafness, 133
Dehydroepiandrosterone sulfate measurement, 36, 43
Developmental delay, 7
Dexamethasone suppression test, 12, 90
Diabetes
    gestational, 45
    insipidus, 76, 197
    mellitus, 61, 121, 151, 197, 225
Diabetic decompensation, 121
Diagnosis, final
    acardic twin sharing womb with normal twin, 131
    acetaminophen poisoning with excessive alcohol intake, 125
    acetone poisoning, 181
    acute hemolytic anemia secondary to systemic lupus erythematosus, 146
    acute intermittent porphyria, 232
    acute phosphate loading caused by hypocalcemia and hypomagnesemia, leading to hypotension due to effects on smooth muscle, 157
    adrenocorticotropic hormone–dependent Cushing's syndrome due to pituitary adrenocorticotropic hormone overproduction with subsequent nodular adrenal hyperplasia, 91
    alcoholic ketoacidosis, 79
    autoimmune vitamin B_{12} deficiency, 31

Burkitt's lymphoma, 159
cerebral salt wasting, 59
chronic hepatitis C infection leading to reduced hepatic synthetic capacity for ceruloplasmin, 153
chronic hyponatremia due to syndrome of inappropriate antidiuretic hormone secretion, 63
congenital adrenal hyperplasia due to 21-hydroxylase deficiency, 16, 21
congenital lactic acidosis due to a pyruvate carboxylase deficiency, 215
Conn's syndrome, 3
Cushing's syndrome caused by a glucocorticoid adenoma, 13
Cushing's syndrome due to pituitary adrenocorticotropic hormone overproduction with subsequent nodular adrenal hyperplasia, 91
cystinosis, 171
DiGeorge syndrome, 115
drug-induced psychosis secondary to leakage of contents when body-packing cocaine, 209
extrahepatic biliary atresia, 99
genetic hemochromatosis, 105
heterozygous familial hypercholesterolemia, 204
hypercalcemia due to primary hyperparathyroidism and/or malignancy, 165
hyperchloremic metabolic acidosis due to transplantation of the ureters into the colon followed by inability of patient to continue taking oral bicarbonate, 193
hyperosmolar diabetic decompensation complicated by rhabdomyolysis, 121
insulinoma, 219
laxative-induced hypokalemia leading to nephrogenic diabetes insipidus, 76
lead poisoning due to ingestion of a traditional Asian remedy containing lead, 111
Leydig cell tumor, 38
lithium-induced nephrogenic diabetes insipidus associated with diabetes mellitus and septicemia, 197
MELAS syndrome, 135
methanol poisoning, 86
multiple endocrine neoplasia type 1, 140

myositis induced by hypolipidemic agents, 204
nephrocalcinosis resulting in acute reversible renal impairment, 157
ornithine carbamoyltransferase deficiency, 223
osteomalacia, 203
ovarian tumor, 38, 43
pancreatitis secondary to hypertriglyceridemia and alcohol abuse, 227
pernicious anemia, 31
pheochromocytoma, 47
polycythemia rubra vera (primary proliferative polycythemia), 67
polyglandular autoimmune syndrome type II with Addison's disease and autoimmune hypothyroidism, 25
primary hyperaldosteronism, 3
pseudoanaphylaxis due to scombroid fish intoxication, 177
raised creatine kinase isoenzyme CK-MB, possibly related to leg weakness or secondary to statin therapy, 95
rhabdomyolysis with hypokalemia, 71
secondary hyperaldosteronism due to congestive heart failure, 9
secondary metabolic bone disease due to long-term therapy with phenytoin, 54
thyrotoxicosis, 31
thyrotropin-secreting tumor concurrent with autoimmune primary hypothyroidism, 149
very-long-chain acyl-coenzyme A dehydrogenase deficiency, 188
Diazoxide therapy, 219
DiGeorge syndrome, 9, 115
1,25-Dihydroxycholecalciferol measurement, 53
1,25-Dihydroxyvitamin D therapy, 171
Diplopia, 133
Diuretics, 9
Dopamine measurement, 46
Dysmorphic facies, 116

E

Edema, 9
Elderly person (≥65 y) as subject of case history, 1, 61, 65, 155, 191, 217
Encephalopathy, 223
Epilepsy, 51, 231

Equations
    anion gap, 84, 181, 193
    corrected sodium value when patient is receiving insulin, 121
    hepatic iron index, 105
    osmolality, calculated, 85
    osmolar gap, 182
Erythrocyte galactose-6-phosphate uridyl transferase measurement, 98
Erythrocyte zinc protoporphyrin, 111
Estradiol measurement, 25, 35, 42, 139
Estriol (unconjugated) measurement, 129
Ethanol
    measurement, 79, 181, 208
    therapy, 85
Ethylene glycol measurement, 182
Extrahepatic biliary atresia, 99

F

Failure to thrive, 169
Ferritin measurement, 98, 104, 152
Ferroxidase activity measurement, 152
α-Fetoprotein measurement, 129
Fish as cause of pseudoanaphylaxis, 177
Fluid input measurement, 58
Fluid output measurement, 58
Folate measurement, 30
Follicle-stimulating hormone measurement, 25, 35, 42, 139
Fomepizole therapy, 85
Foreign travel, 207
Free androgen index calculation, 35

G

Gastrin measurement, 139
Genetic hemochromatosis, 105
Globulins measurement, 145
Glucocorticoid adenoma, 13
Glucose
    measurement, 8, 24, 30, 45, 62, 73, 78, 83, 103, 110, 120, 134, 138, 175, 180, 192, 195, 202, 213, 217, 225
    therapy, 80
    tolerance test, 45

γ-Glutamyltransferase measurement, 53, 78, 83, 95, 98, 130, 202, 218, 222, 226
Glycoprotein hormone α-subunit, 149
Growth disorder, 19
Growth hormone measurement, 139, 218

H

Heavy metals in Asian cosmetics and remedies, 112
Hematological parameter measurement, 24, 30, 46, 66, 69, 78, 83, 98, 110, 125, 138, 143, 151, 170, 180, 186, 192, 208, 213, 222, 226
Hemochromatosis, 105
Hemoglobin A_{1c}, 151
Hemolytic anemia, 146
Heparin as cause of hypoaldosteronism, 135
Hepatitis C
 infection, 153
 screen, 98, 110, 151
Heroin measurement, 208
Heterozygous familial hypercholesterolemia, 204
HFE gene, 105
Hirsutism, 35, 89
HMBS gene, 233
Homovanillic acid measurement, 46
Human choriogonadotropin, 129
Hydrocortisone therapy, 16, 21, 25
β-Hydroxybutyrate, 79
25-Hydroxycholecalciferol measurement, 203
21-Hydroxylase deficiency, 16, 21
4-Hydroxy-3-methoxymandelic acid measurement, 46
3-Hydroxy-3-methylglutaryl coenzyme A reductase inhibitor therapy, 201
17α-Hydroxyprogesterone measurement, 16, 21, 43
$α_1$-Hydroxyvitamin D therapy, 115
25-Hydroxyvitamin D measurement, 114
Hyperaldosteronism, 3, 9
Hyperbilirubinemia, 100
Hypercalcemia
 caused by primary hyperparathyroidism and/or malignancy, 165
 classification of tumor hypercalcemia, 166
Hyperchloremic metabolic acidosis, 193
Hypercholesterolemia, 93, 204
Hypercortisolemia of pregnancy, 13
Hypergastrinemia, 140
Hyperhidrosis, 47
Hyperkalemia, 65, 135
Hyperlipidemia, some causes, *228*
Hypernatremia, 121
Hyperosmolar diabetic decompensation, 121
Hyperparathyroidism, 140, 165
Hyperprolactinemia, 140
Hypertension, 1, 45, 70, 157
Hypertriglyceridemia, 145, 227
Hypocalcemia, 115, 157
Hypoglycemia, 217
Hypokalemia, 70, 76
Hypomagnesemia, 157
Hyponatremia, 9, 57, 61
Hypotension, 23, 157
Hypothermia, 179
Hypothyroidism, 25
Hypouricemia, 160

I

Immunoglobulin A measurement, 151
Immunoglobulin E measurement, 176
Immunoglobulin G measurement, 151
Immunoglobulin M measurement, 151
Immunoreactive trypsin measurement, 98
Infant (<1 y) as subject of case history, 15, 41, 97, 113, 213
Infertility, 29, 35, 109
Insulin measurement, 218
Insulinoma, 219
Intestinal peptide screen, 219
Iron-binding capacity (total), 104
Iron
 disorders of metabolism, 106, 153
 measurement, 30, 104, 152
 overload, 105
Ischemic exercise test, 186
Isopropanol
 measurement, 181
 toxicity, 182

J

Jaundice, 97

K

Ketoacidosis caused by alcoholism, 79
Ketone measurement, 79, 120, 180, 196, 218

L

Lactate dehydrogenase, 95
Lactate
    measurement, 7, 79, 186, 214
    ratio to pyruvate, 186
Lactic acid measurement, 181
Laxative-induced hypokalemia, 76
Lead
    effects on heme synthesis, *111*
    high concentrations in Asian cosmetics and remedies, 112
    measurement, 111
    poisoning, 111
Leg pain, 71, 103
Leydig cell tumor, 38
Lithium
    as cause of nephrogenic diabetes insipidus, 197
    measurement, 195
    therapy, 198
Loss of consciousness, 57, 83, 133, 179, 213, 217, 222
Luteinizing hormone measurement, 25, 35, 42, 139

M

Macrocytosis, 30
Magnesium
    measurement, 155, 159, 170, 203
    therapy, 171
Melanocyte-stimulating hormone, 27
MELAS syndrome, 135
Metabolic acidosis. *See* Acidosis
Metabolic alkalosis. *See* Alkalosis
Metabolic bone disease, 54
Methanol
    measurement, 86, 181
    metabolism, *86*
    poisoning, 86
Moon facies, 89
Morphine measurement, 208
Multiple endocrine neoplasia type 1, 140
Multiples of the median calculation, 129

Muscle
    pain, 51, 201
    wasting, 169
    weakness, 204
Muscle acyl coenzyme A (CoA) dehydrogenase specific activity
    measurement of octanoyl CoA, 187
    measurement of palmitoyl CoA, 187
    measurement of oleoyl CoA, 187
Muscle mitochondrial respiratory chain complexes
    ratio of complex I to citrate synthase, 187
    ratio of complex II/III to citrate synthase, 187
    ratio of complex IV to citrate synthase, 187
Myoglobin measurement, 95
Myositis, drug induced, 204

N

Natriuresis, 59
Neonatal hypocalcemia, causes, 115
Nephrocalcinosis, 157
Nephrogenic diabetes insipidus, 76, 197
Nocturia, 217
Noradrenaline measurement, 46

O

Octreotide therapy, 219
Ornithine carbamoyltransferase deficiency, 223
Orotate-creatinine ratio calculation, 223
Osmolality
    equation for calculating corrected sodium value when patient is receiving insulin, 121
    measurement, 8, 58, 62, 71, 75, 84, 120, 180, 192, 195
Osmolar gap calculation, 85
Osteomalacia, 204
Ovarian tumor, 38, 43

P

Pancreatitis, 227
Paracetamol. *See* Acetaminophen
Parathyroid hormone measurement, 47, 53, 114, 139, 156, 165, 203, 219
Pernicious anemia, 31

Phenytoin
    as cause of bone loss, 54
    measurement, 52
Pheochromocytoma, 47
Phosphate
    loading (accidental iatrogenic), 157
    measurement, 52, 113, 144, 155, 159, 164, 170, 203, 218
Phosphoenolpyruvate carboxykinase activity in cultured fibroblasts measurement, 215
Pituitary
    adenoma, 149
    function tests, 139, 149
Polycythemia
    primary proliferative form, 67
    secondary and relative causes, 68
Polydipsia, 70, 119, 195
Polyglandular autoimmune syndrome type II, 25
Polyuria, 59, 70, 119, 195
Porphobilinogen measurement, 232
Porphobilinogen deaminase measurement, 232
Porphyria, 233
Porphyrin measurement, 232
Potassium
    measurement, 3, 8, 20, 24, 52, 58, 61, 66, 69, 73, 78, 83, 90, 103, 110, 120, 124, 130, 134, 138, 155, 159, 164, 170, 175, 180, 192, 195, 202, 208, 218, 226
    output measurement, 74
    therapy, 71, 76, 80, 157
Precocious development, 43
    See also Precocious puberty
Precocious puberty, 21
Pregnancy
    Cushing's syndrome in, 13
    gestational diabetes, 45
    hypercortisolemia of pregnancy, 13
    hypokalemia in, 70
    insulin resistance in, 48
    second-trimester screening, 129
    twins, 129
    use of carbimazole, 33
Pregnant woman as subject of case history, 11, 69, 129
Primary hyperaldosteronism, 3
Prolactin measurement, 25, 42, 139, 149
Propylthiouracil therapy, 31

Protein measurement
    in cerebrospinal fluid, 134
    ratio to creatinine, 171
    total, 11, 52, 66, 69, 98, 104, 130, 138, 144, 202
Pseudoanaphylaxis due to scombroid fish intoxication, 177
Psychosis induced by cocaine, 209
Pyruvate carboxylase
    activity in cultured fibroblasts measurement, 215
    deficiency, 215
Pyruvate dehydrogenase activity in cultured fibroblasts measurement, 215

R

Renal failure, 185
Renin-angiotensin system, *4*
Renin activity measurement, 3, 8, 20, 62, 74, 135, 171
Rhabdomyolysis, 71, 121, 185
Rheumatoid factor measurement, 103

S

Salicylate measurement, 84, 124, 180
Salt-losing crisis, 16
Scombroid fish intoxication, 177
Secondary hyperaldosteronism, 9
Seizures, 207, 221, 231
Septicemia, 197
Sex hormone–binding globulin measurement, 35
Skin pigmentation, 27
Skin prick test for allergies, 176
Sodium
    equation for calculating corrected value when patient is receiving insulin, 121
    measurement, 3, 8, 20, 24, 52, 57, 61, 66, 69, 73, 78, 83, 103, 110, 120, 124, 130, 134, 138, 155, 159, 164, 170, 175, 180, 192, 195, 202, 208, 218, 226
    therapy, 59
Steroid hormone synthesis, 17, *22*
Stroke, 133
Subarachnoid hemorrhage, 57
Sulfonylurea measurement, 219
Synacthen test, 25, 62
Syndrome of inappropriate antidiuretic hormone secretion, 58, 63
Systemic lupus erythematosus, 146

T

Tachycardia, 29
Tachypnea, 77, 179
Testicular feminization syndrome, 37
Testosterone measurement, 20, 35, 43
Thyroid microsomal antibody measurement, 30
Thyroid-stimulating hormone. *See* Thyrotropin (thyroid-stimulating hormone)
Thyrotoxicosis, 31, 147
Thyrotropin (thyroid-stimulating hormone)
    measurement, 25, 30, 46, 58, 62, 95, 98, 103, 139, 147, 202
    secretion by tumor, 149
Thyrotropin-releasing hormone–gonadotropin hormone–releasing hormone test, 139, 149
Thyroxine
    measurement, 25, 30, 46, 58, 62, 98, 103, 139, 147, 202
    therapy, 25, 149, 165
Tingling and swelling of lips, 175
Transferrin
    measurement, 152
    saturation measurement, 104
Triglyceride measurement, 98, 144, 202, 226
Triiodothyronine measurement, 30, 58, 147
Troponin I measurement, 94
Tumor
    adrenocorticotropic hormone secreting, 91
    estrogen secreting, 43
    hypercalcemia classification, 166
    in the sigmoid colon, 163
    pheochromocytoma, 47
    testosterone secreting, 38
    thyrotropin secreting, 149
Tumor lysis syndrome, 160
Twin, acardic, 131

U

Urate measurement, 159, 202
Urea measurement, 3, 8, 20, 24, 52, 62, 66, 69, 73, 78, 83, 95, 103, 110, 120, 124, 130, 134, 138, 144, 155, 159, 170, 175, 180, 192, 195, 202, 208, 226
Uric acid measurement, 103, 186
Uricozyme (urate oxidase) therapy, 161
Urinary steroid profile, 16
Urine
    dark, 185, 231
    pink, 75
    red, 120
Urine volume
    in relation to urine osmolality, *64*
    measurement, 63, 74
Ursodeoxycholic acid therapy, 99

V

Valproate therapy, 223
Vasopressin (antidiuretic hormone), 198
Venesection therapy, 105
Very-long-chain acyl-coenzyme A dehydrogenase deficiency, 188
Viral screen, 98, 103, 130, 186
    See also Hepatitis C screen
Virilization, 16, 21
Vitamin A therapy, 99
Vitamin B_{12}
    absorption, *32*
    deficiency caused by autoimmunity, 31
    measurement, 30
    therapy, 31
Vitamin D
    therapy, 54, 99, 115, 140
    See also 1,25-Dihydroxycholecalciferol measurement; 1,25-Dihydroxyvitamin D therapy; 25-Hydroxycholecalciferol measurement; α_1-Hydroxyvitamin D therapy; 25-Hydroxyvitamin D measurement
Vitamin E therapy, 99
Vitamin K therapy, 99

W

Water
    deprivation test, 75, 196
    load-deprivation test, 63